Astronomy FOR DUMMIES®

2ND EDITION

by Stephen P. Maran, PhD

WILEY

Wiley Publishing, Inc.

Astronomy For Dummies®, 2nd Edition

Published by
Wiley Publishing, Inc.
111 River St.
Hoboken, NJ 07030-5774
www.wiley.com

For general information on our other products and services, please contact our Customer Care Department within the U.S. at 800-762-2974, outside the U.S. at 317-572-3993, or fax 317-572-4002.

For technical support, please visit www.wiley.com/techsupport.

Wiley also publishes its books in a variety of electronic formats. Some content that appears in print may not be available in electronic books.

Library of Congress Control Number: 2005923245

ISBN-13: 978-0-7645-8465-7

ISBN-10: 0-7645-8465-0

Manufactured in the United States of America

10 9 8 7 6 5

2O/SQ/QU/QV/IN

WILEY

About the Author

Stephen P. Maran, PhD, a 36-year veteran of the space program, was the 1999 recipient of the Klumpke-Roberts Award of the Astronomical Society of the Pacific for "outstanding contributions to the public understanding and appreciation of astronomy." He received the NASA Medal for Exceptional Achievement in 1991 and was the A. Dixon Johnson Lecturer in Scientific Communication at Pennsylvania State University in 1990. In March of 2000, the International Astronomical Union named an asteroid (Minor Planet 9768) "Stephenmaran" in his honor. He's also taught astronomy at the University of California, Los Angeles and the University of Maryland, College Park. As press officer of the American Astronomical Society, he presides over media briefings that bring the news of astronomical discoveries to people worldwide.

Dr. Maran began practicing astronomy from rooftops in Brooklyn and at a deserted golf course in the far reaches of the Bronx. He graduated to conduct professional research with telescopes at the Kitt Peak National Observatory in Arizona, the National Radio Astronomy Observatory in West Virginia, the Palomar Observatory in California, and the Cerro Tololo Inter-American Observatory in Chile. He also conducted research with instruments in space, including the Hubble Space Telescope and the International Ultraviolet Explorer. He helped design and develop two instruments that flew in space aboard Hubble. He's observed total eclipses of the sun from the Gaspé Peninsula and elsewhere in Quebec, from Baja California in Mexico, at sea off New Caledonia and Singapore, in the eastern Pacific, and from the shores of the good old U.S.A.

In the course of spreading the good word about astronomy, Maran has lectured on black holes in a bar in Tahiti and explained an eclipse of the sun on NBC's *Today* show. He's also spoken on eclipse and comet cruises aboard Cunard's *Queen Elizabeth 2* and *Vistafjord* and Sitmar Line's *Fairwind.* He's addressed audiences ranging from Seattle school children and Atherton, California Girl Scouts to the National Academy of Engineering in Washington, D.C. and subcommittees of both the U.S. House of Representatives and the U.N. Committee on the Peaceful Uses of Outer Space.

Dr. Maran is editor of *The Astronomy and Astrophysics Encyclopedia* and has co-authored or edited eight other books on those subjects, including a college textbook, *New Horizons in Astronomy,* and two compendiums of space program discoveries, *A Meeting with the Universe* and *Gems of Hubble.* He has written many articles for *Smithsonian* and *Natural History* magazines and has served as a writer and consultant for the National Geographic Society and Time-Life Books.

Dr. Maran is a graduate of Stuyvesant High School in New York City, where he served a full season on the Math Team without incurring serious injury, and of Brooklyn College. He received both his MA and PhD in astronomy from the University of Michigan. Maran is married to Sally Scott Maran, a journalist. They have three children.

Dedication

To Sally, Michael, Enid, and Elissa with all my love.

Author's Acknowledgments

Thanks first to my family and friends who put up with me in the writing of this book. Thanks also to my agent, Skip Barker of the Wilson-Devereaux Company, who goaded and guided me in this project, and to Stacy Collins for her faith in the original project.

I'm grateful to Ron Cowen and Dr. Seth Shostak for their contributions to this book; to Kathy Cox, Georgette Beatty, and Josh Dials, who organized and edited it; and to their skilled colleagues on the editorial and production teams at Wiley Publishing who made the book better and brighter. A special thanks to Dr. Matthew Lister of Purdue University for suggestions that improved the accuracy and thoroughness of the book.

Thanks as well to the organizations that provide the photographs in this book; to the producer of the star maps, Robert Miller; and to the producer of the planetary tables, Martin Ratcliffe.

Some drawings in this book may have been inspired by Dr. Dinah L. Moche and her excellent book, *Astronomy, A Self-Teaching Guide,* also published by Wiley. Dr. Moche deserves many thanks for her support of this book and for her dedication to making the science of astronomy accessible to anyone.

Publisher's Acknowledgments

We're proud of this book; please send us your comments through our Dummies online registration form located at www.dummies.com/register/.

Some of the people who helped bring this book to market include the following:

Acquisitions, Editorial, and Media Development

Project Editor: Georgette Beatty

(Previous Edition: Kathy Cox)

Acquisitions Editor: Kathy Cox

Copy Editor: Josh Dials

(Previous Edition: Susan Diane Smith)

Technical Editor: Matthew Lister

Editorial Manager: Michelle Hacker

Editorial Assistants: Hanna Scott, Nadine Bell

Cover Photo: Courtesy of NASA

Cartoons: Rich Tennant (www.the5thwave.com)

Composition Services

Project Coordinator: Nancee Reeves

Layout and Graphics: Andrea Dahl, Joyce Haughey, Shelley Norris, Barry Offringa, Lynsey Osborn, Melanee Prendergast, Heather Ryan

Proofreaders: Laura Albert, Leeann Harney, Jessica Kramer, TECHBOOKS Production Services

Indexer: TECHBOOKS Production Services

Publishing and Editorial for Consumer Dummies

> **Diane Graves Steele,** Vice President and Publisher, Consumer Dummies
>
> **Joyce Pepple,** Acquisitions Director, Consumer Dummies
>
> **Kristin A. Cocks,** Product Development Director, Consumer Dummies
>
> **Michael Spring,** Vice President and Publisher, Travel
>
> **Kelly Regan,** Editorial Director, Travel

Publishing for Technology Dummies

> **Andy Cummings,** Vice President and Publisher, Dummies Technology/General User

Composition Services

> **Gerry Fahey,** Vice President of Production Services
>
> **Debbie Stailey,** Director of Composition Services

Contents at a Glance

Introduction 1

Part I: Stalking the Cosmos 7

Chapter 1: Seeing the Light: The Art and Science of Astronomy 9
Chapter 2: Join the Crowd: Skywatching Activities and Resources 29
Chapter 3: The Way You Watch Tonight: Terrific Tools for Observing the Skies 41
Chapter 4: Just Passing Through: Meteors, Comets, and Artificial Satellites 57

Part II: Going Once Around the Solar System 75

Chapter 5: A Matched Pair: Earth and Its Moon 77
Chapter 6: Earth's Near Neighbors: Mercury, Venus, and Mars 97
Chapter 7: Rock On: The Asteroid Belt and Near-Earth Objects 115
Chapter 8: Great Balls of Gas: Jupiter and Saturn 123
Chapter 9: Far Out! Uranus, Neptune, Pluto, and Beyond 135

Part III: Meeting Old Sol and Other Stars 143

Chapter 10: The Sun: Star of the Earth 145
Chapter 11: Taking a Trip to the Stars 169
Chapter 12: Galaxies: The Milky Way and Beyond 197
Chapter 13: Digging into Black Holes and Quasars 219

Part IV: Pondering the Remarkable Universe 231

Chapter 14: Is Anybody Out There? SETI and Planets of Other Suns 233
Chapter 15: Delving into Dark Matter and Antimatter 245
Chapter 16: The Big Bang and the Evolution of the Universe 253

Part V: The Part of Tens 263

Chapter 17: Ten Strange Facts about Astronomy and Space 265
Chapter 18: Ten Common Errors about Astronomy and Space 269

Part VI: Appendixes 273

Appendix A: Finding the Planets: 2006 to 2010 275
Appendix B: Star Maps 293
Appendix C: Glossary 301

Index 305

Table of Contents

Introduction *1*

About This Book 2
Conventions Used in This Book 2
What You're Not to Read 2
Foolish Assumptions 3
How This Book Is Organized 3
    Part I: Stalking the Cosmos 3
    Part II: Going Once Around the Solar System 4
    Part III: Meeting Old Sol and Other Stars 4
    Part IV: Pondering the Remarkable Universe 4
    Part V: The Part of Tens 4
    Part VI: Appendixes 5
Icons Used in This Book 5
Where to Go from Here 5

Part 1: Stalking the Cosmos *7*

Chapter 1: Seeing the Light: The Art and Science of Astronomy 9

Astronomy: The Science of Observation 10
Understanding What You See: The Language of Light 11
    They wondered as they wandered: Planets versus stars 12
    If you see a Great Bear, start worrying: Naming stars
      and constellations 12
    What do I spy? The Messier Catalog and other sky objects 20
    The smaller, the brighter: Getting to the root of magnitudes 20
    Looking back on light-years 22
    Keep on moving: Figuring the positions of the stars 23
Gravity: A Force to Be Reckoned With 26
Space: A Commotion of Motion 27

Chapter 2: Join the Crowd: Skywatching Activities
and Resources 29

You're Not Alone: Astronomy Clubs, Web Sites, and More 30
    Joining an astronomy club for star-studded company 30
    Checking Web sites, magazines, and software 31
Visiting Observatories and Planetariums 34
    Ogling the observatories 34
    Popping in on planetariums 36

Vacationing with the Stars: Star Parties, Eclipse Trips,
   and Telescope Motels 36
      Party on! Attending star parties 37
      To the path of totality: Taking eclipse cruises and tours 38
      Motoring to telescope motels 40

**Chapter 3: The Way You Watch Tonight: Terrific Tools
for Observing the Skies** **41**
   Seeing Stars: A Sky Geography Primer 42
      As the earth turns 42
      . . . keep an eye on the North Star 43
   Beginning with Naked-Eye Observation 45
   Using Binoculars or a Telescope for a Better View 48
      Binoculars: Sweeping the night sky 48
      Telescopes: When closeness counts 50
   Planning Out Your Dip into Astronomy 56

**Chapter 4: Just Passing Through: Meteors, Comets,
and Artificial Satellites** **57**
   Meteors: Wishing on a Shooting Star 57
      Spotting sporadic meteors, fireballs, and bolides 59
      Watching a radiant sight: Meteor showers 61
   Comets: The Lowdown on Dirty Ice Balls 65
      Making heads and tails of a comet's structure 65
      Waiting for the "comets of the century" 68
      Hunting for the great comet 70
   Artificial Satellites: Enduring a Love-Hate Relationship 72
      Skywatching for artificial satellites 73
      Finding satellite viewing predictions 74

Part II: Going Once Around the Solar System *75*

Chapter 5: A Matched Pair: Earth and Its Moon **77**
   Putting Earth Under the Astronomical Microscope 78
      One of a kind: Earth's unique characteristics 78
      Spheres of influence: Earth's distinct regions 80
   Examining Earth's Time, Seasons, and Age 82
      Orbiting for all time 82
      Tilting toward the seasons 84
      Estimating Earth's age 86
   Making Sense of the Moon 87
      Get ready to howl: Phases of the moon 87
      In the shadows: Watching lunar eclipses 89
      Hard rock: Viewing the moon's geology 90
      Quite an impact: A theory about the moon's origin 94

Chapter 6: Earth's Near Neighbors: Mercury, Venus, and Mars 97

Hot, Shrunken, and Battered: Putting Mercury on a Platter 97
Dry, Acidic, and Hilly: Steering Clear of Venus 99
Red, Cold, and Barren: Uncovering the Mysteries of Mars 100
　Where has all the water gone? 101
　Does Mars support life? 102
Differentiating Earth through Comparative Planetology 103
Observing the Terrestrial Planets with Ease 104
　Understanding elongation, opposition, and conjunction 106
　Viewing Venus and its phases 108
　Watching Mars as it loops around 110
　Outdoing Copernicus by observing Mercury 113

Chapter 7: Rock On: The Asteroid Belt and Near-Earth Objects 115

Taking a Brief Tour of the Asteroid Belt 115
Understanding the Threat That Near-Earth Objects Pose 118
　When push comes to shove: Nudging an asteroid 119
　Forewarned is forearmed: Surveying NEOs to protect Earth 119
Searching for Small Points of Light 120
　Timing an asteroidal occultation 121
　Helping to track an occultation 122

Chapter 8: Great Balls of Gas: Jupiter and Saturn 123

The Pressure's On: Journeying Inside Jupiter and Saturn 123
Almost a Star: Gazing at Jupiter 124
　Scanning for the Great Red Spot 126
　Shooting for Galileo's moons 127
Our Main Planetary Attraction: Setting Your Sights on Saturn 130
　Ringing around the planet 130
　Stormchasing across Saturn 132
　Monitoring a moon of major proportions 132

Chapter 9: Far Out! Uranus, Neptune, Pluto, and Beyond 135

Breaking the Ice with Uranus and Neptune 135
　Bull's-eye! Tilted Uranus and its features 136
　Against the grain: Neptune and its moon 137
Meeting Pluto, an Unusual Planet 137
　The moon chip doesn't float far from the planet 138
　The little planet with little respect 139
Buckling Down to the Kuiper Belt 139
Viewing the Outer Planets 140
　Sighting Uranus 141
　Distinguishing Neptune from a star 141
　Straining to see Pluto 142

Part III: Meeting Old Sol and Other Stars 143

Chapter 10: The Sun: Star of the Earth 145
Surveying the Sunscape 146
    The sun's size and shape: A great bundle of gas 147
    The sun's regions: Caught between the core and the corona 147
    Solar activity: What's going on out there? 149
    Solar wind: Playing with magnets 153
    Solar CSI: The mystery of the missing solar neutrinos 153
    Four billion and counting: The life expectancy of the sun 154
Don't Make a Blinding Mistake: Safe Techniques for Solar Viewing 155
    Viewing the sun by projection 156
    Viewing the sun through front-end filters 159
Fun with the Sun: Solar Observation 160
    Tracking sunspots 160
    Experiencing solar eclipses 162
    Looking at solar pictures on the Net 167

Chapter 11: Taking a Trip to the Stars 169
Life Cycles of the Hot and Massive 169
    Young stellar objects: Taking baby steps 171
    Main sequence stars: A long adulthood 172
    Red giants: Burning out the golden years 172
    Closing time: Stars at the tail end of stellar evolution 173
Diagramming Star Color, Brightness, and Mass 178
    Spectral types: What color is my star? 178
    Star light, star bright: Classifying luminosity 180
    The brighter they burn, the bigger they swell: Mass
        determines class 181
    Interpreting the H-R diagram 182
Eternal Partners: Binary and Multiple Stars 183
    Binary stars and the Doppler Effect 183
    Two stars are binary, but three's a crowd: Multiple stars 187
Change Is Good: Variable Stars 187
    Going the distance: Pulsating stars 188
    Explosive neighbors: Flare stars 190
    Nice to nova: Exploding stars 190
    Stellar hide and seek: Eclipsing binary stars 192
    Hogging the starlight: Microlensing events 193
Meeting Your Stellar Neighbors 193
Helping Scientists by Observing the Stars 195

Chapter 12: Galaxies: The Milky Way and Beyond 197
Unwrapping the Milky Way 197
    How and when did the Milky Way form? 198
    What shape is the Milky Way? 199
    Where can you find the Milky Way? 200

Star Clusters: Galactic Associates 201
    A loose fit: Open clusters 202
    A tight squeeze: Globular clusters 203
    Fun while it lasted: OB associations 205
Taking a Shine to Nebulae 205
    Picking out planetary nebulae 206
    Breezing through supernova remnants 208
    Enjoying Earth's best nebular views 208
Getting a Grip on Galaxies 211
    Surveying spiral, barred spiral, and lenticular galaxies 211
    Examining elliptical galaxies 212
    Looking at irregular, dwarf, and low surface brightness
        galaxies 213
    Gawking at great galaxies 214
    Discovering the Local Group of Galaxies 216
    Checking out clusters of galaxies 217
    Sizing up superclusters, cosmic voids, and Great Walls 218

Chapter 13: Digging into Black Holes and Quasars **219**
Black Holes: There Goes the Neighborhood 219
    Looking over the black-hole roster 220
    Poking around the black-hole interior 220
    Surveying a black hole's surroundings 223
    Warping space and time 224
Quasars: Defying Definitions 225
    Measuring the size of a quasar 226
    Getting up to speed on jets 226
    Exploring quasar spectra 226
Active Galactic Nuclei: Welcome to the Quasar Family 227
    Sifting through different types of AGN 227
    Examining the power behind AGN 229
    Proposing the Unified Model of AGN 230

Part 1V: Pondering the Remarkable Universe *231*

**Chapter 14: Is Anybody Out There? SETI and Planets
of Other Suns** **233**
Using Drake's Equation to Discuss SETI 234
SETI Projects: Listening for E.T. 235
    The flight of Project Phoenix 237
    Space scanning with other SETI projects 238
    SETI programs want you! 240
Finding Extrasolar Planets 240
    51 Pegasi's hot partner 242
    The Upsilon Andromedae system 243
    Continuing the search for planets suitable for life 243

Chapter 15: Delving into Dark Matter and Antimatter 245

Dark Matter: Understanding the Universal Glue 245
    Gathering the evidence for dark matter 246
    Debating the makeup of dark matter 248
Taking a Shot in the Dark: Searching for Dark Matter 249
    WIMPs: Leaving a weak mark 249
    MACHOs: Making a brighter image 250
    Mapping dark matter with gravitational lensing 250
Dueling Antimatter: Proving That Opposites Attract 251

Chapter 16: The Big Bang and the Evolution of the Universe 253

Assessing Evidence for the Big Bang 254
Inflation: A Swell Time in the Universe 255
    Something from nothing: Inflation and the vacuum 256
    Falling flat: Inflation and the shape of the universe 257
Dark Energy: Stepping on the Universal Accelerator 258
Pulling Universal Info from the Cosmic Microwave Background 258
    Finding the lumps in the cosmic microwave background 259
    Mapping the universe with the cosmic microwave
    background 260
In a Galaxy Far Away: The Hubble Constant and Standard Candles 260
    The Hubble constant: How fast do galaxies really move? 261
    Standard candles: How do scientists measure galaxy
    distances? 262

Part V: The Part of Tens 263

Chapter 17: Ten Strange Facts about Astronomy and Space 265

You Have Tiny Meteorites in Your Hair 265
A Comet's Tail Often Leads the Way 265
Earth Is Made of Rare and Unusual Matter 266
High Tide Comes on Both Sides of the Earth at the Same Time 266
On Venus, the Rain Never Falls on the Plain 266
Rocks from Mars Dot the Earth 266
Pluto Was Discovered from the Predictions of a False Theory 267
Sunspots Are Not Dark 267
A Star in Plain View May Have Exploded, but No One Knows 267
You May Have Seen the Big Bang on an Old Television 268

Chapter 18: Ten Common Errors about Astronomy and Space 269

"The Light from That Star Took 1,000 Light-Years to Reach Earth" 269
A Freshly Fallen Meteorite Is Still Hot 269
Summer Always Comes When Earth Is Closest to the Sun 270

The "Morning Star" Is a Star 270
If You Vacation in the Asteroid Belt, You'll See Asteroids
    All Around You 270
Nuking a "Killer Asteroid" on a Collision Course for Earth Will
    Save Us 270
Asteroids Are Round, Like Little Planets 271
The Sun Is an Average Star 271
The Hubble Telescope Gets Up Close and Personal 271
The Big Bang Is Dead 271

Part VI: Appendixes *273*

Appendix A: Finding the Planets: 2006 to 2010 ....**275**
2006 276
2007 279
2008 282
2009 285
2010 288

Appendix B: Star Maps **293**

Appendix C: Glossary **301**

Index *305*

Introduction

*A*stronomy is the study of the sky, the science of cosmic objects and celestial happenings. It's nothing less than the investigation of the nature of the universe we live in. Astronomers carry out the business of astronomy by looking and (for radio astronomers) listening. Astronomy is done with backyard telescopes, huge observatory instruments, and satellites orbiting Earth or positioned in space near Earth or another celestial body, such as the moon or a planet. Scientists send up telescopes in sounding rockets and on unmanned balloons; some instruments travel far into the solar system aboard deep space probes; and some probes gather samples with the aim of returning them to Earth.

Astronomy can be a professional or an amateur activity. About 15,000 professional astronomers engage in space science worldwide. Over 300,000 amateur astronomers live in the United States alone. And amateur astronomy clubs are everywhere.

Professional astronomers conduct research on the sun and the solar system, the Milky Way galaxy, and the universe beyond. They teach in universities, design satellites in government labs, and operate planetariums. They also write books, like this one (but maybe not as good). Most hold PhDs, and nowadays — so many of them study abstruse physics or work with automated, robotic telescopes — they may not even know the constellations.

Amateur astronomers know the constellations. They share an exciting hobby. Some stargaze on their own, and thousands more join astronomy clubs and organizations of every description. The clubs pass on know-how from old hands to new members, share telescopes and equipment, and hold meetings where members tell about their recent observations or hear lectures by visiting scientists.

Amateur astronomers also hold observing meetings where everyone brings a telescope (or looks through another observer's scope). The amateurs conduct these sessions at regular intervals (such as the first Saturday night of each month) or on special occasions (such as the return of a major meteor shower each August or the appearance of a bright comet like Hale-Bopp). And they save up for really big events, such as a total eclipse of the sun, when thousands of amateurs and dozens of pros travel across Earth to position themselves in the path of totality to witness one of nature's greatest spectacles.

About This Book

This book explains all you need to know to launch into the great hobby of astronomy. And it gives you a leg up on understanding the basic science of the universe as well. The latest space missions will make more sense: You'll understand why NASA and other organizations send space probes to planets like Saturn, why robot rovers land on Mars, and why scientists seek samples of the dust in the tail of a comet. You'll know why the Hubble Space Telescope peers out into space and how to check up on other space missions. And when astronomers show up in the newspaper or on television to report their latest discoveries — from space; from the big telescopes in Arizona, Hawaii, Chile, and California; or from other observatories around the world — you'll understand the background and appreciate the news. You can even explain it to your friends.

Read only the parts you want, in any order you want. I explain what you need as you go. Astronomy is fascinating and fun, so keep reading. Before you know it, you'll be pointing out Jupiter, spotting famous constellations and stars, and tracking the International Space Station as it whizzes by overhead. The neighbors may start calling you "stargazer." Police officers may ask you what you're doing in the park at night or why you're standing on the roof with binoculars. Tell 'em you're an astronomer. That's one they probably haven't heard (I hope they believe you!).

Conventions Used in This Book

To help you navigate this book as you begin to navigate the skies, I use the following conventions:

- *Italic* text highlights new words and defined terms.
- **Boldfaced** text indicates keywords in bulleted lists and the action part of numbered steps.
- Monofont text highlights a Web address.

What You're Not to Read

Feel free to skip the sidebars that appear throughout the book; these shaded gray boxes contain interesting info that isn't essential to your understanding of astronomy. The same goes for any text I mark with the Technical Stuff icon.

Foolish Assumptions

You may be reading this book because you want to know what's up in the sky or what the scientists in the space program are doing. Perhaps you've heard that astronomy is a neat hobby, and you want to find out if the rumor is true. Perhaps you want to find out what equipment you need.

You're not a scientist. You just enjoy looking at the night sky and have fallen under its spell, wanting to see and understand the real beauty of the universe.

You want to observe the stars, but you also want to know what you're seeing. Maybe you even want to make a discovery of your own. You don't have to be an astronomer to spot a new comet, and you can even help listen for E.T. Whatever your goal, this book helps you achieve it.

How This Book Is Organized

If you've already peeked at the table of contents, you know I divide this book into parts. Here's a brief description of what you can find in each of the six major parts.

Part 1: Stalking the Cosmos

You see the stars night after night (well, not every night, but still . . .). You develop the same fascination that humans have always had with the cosmos. You watch, and you wonder, and you want to know more. What are those lights in the sky? What makes them look the way they do and move the way they move? Are any of the objects dangerous? Should you be waving at your cosmic twin?

This part sends you down the path toward answers to some of these questions, based on answers that astronomers have already found. Thousands of amateur astronomers gather together to support each other and share their knowledge. Astronomy is fun, as well as practical (and, yes, even educational).

In this part, I give you pointers on observing sky objects with and without optics, selecting binoculars and telescopes, and positioning yourself for the best view. I introduce you to delightful cosmic visitors, and set you up to continue exploring the universe with the help of various resources.

Part II: Going Once Around the Solar System

It's only natural to want to meet your neighbors. Earth's neighbors are a collection of planets, moons, and planetary debris linked by the circumstance that they all orbit around the sun. Like all neighbors, these solar system objects share certain characteristics, but they're all wildly different, too.

These chapters focus on the observational aspects of the planets to help you know what you're seeing so you can appreciate the view.

Part III: Meeting Old Sol and Other Stars

Wondering about galaxies far, far away? This part starts with the sun and takes you through the stars. It introduces you to red giants and white dwarfs, drops in on distant galaxies and exotic sky objects, and ends with black holes. Do you really want to go there? You may get swept away.

As the late, great Carl Sagan said, "We are all star stuff." So understanding the stars and enjoying their diversity enhances your connections to the stuff of the universe.

This part points out the best and the brightest of the sky objects for your observational pleasure. I also spell out the life cycles of the stars so you can appreciate the forces that power the universe and make it endlessly intriguing.

Part IV: Pondering the Remarkable Universe

This part is all about thought-provoking concepts like the search for extraterrestrial intelligence and the nature of dark matter and antimatter. It even describes concepts of the universe as a whole: its beginnings, its current shape, and its likely fate.

Part V: The Part of Tens

This part offers 10 strange facts about space and 10 mistakes that the public and the media often make when talking about astronomy. Read on to avoid making these mistakes.

Part VI: Appendixes

This part provides appendixes that help to enhance your skywatching for many years to come. The first appendix presents tables of the positions of the four brightest planets — Venus, Mars, Jupiter, and Saturn — so you can easily find them season by season and year by year. The second offers star maps to help you discover the constellations. And the third is a glossary of common terms in astronomy.

Icons Used in This Book

Throughout this book, helpful icons highlight particularly useful information — even if they just tell you to not sweat the tough stuff. Here's what each symbol means.

Observation is the key to astronomy, and these tips help make you an observational pro. I help you scope out techniques and opportunities to fine-tune your viewing technique.

This nerdy guy appears beside discussions that you can skip if you just want to know the basics and start watching the skies. The scientific background can be good to know, but many people happily enjoy their stargazing without knowing about the physics of supernovas, the mathematics of galaxy chasing, and the ins and outs of dark energy.

This target puts you right on track to make use of some inside information as you start skywatching or make progress in the hobby.

How much trouble can you get into watching the stars? Not much, if you're careful. But some things you can't be too careful about. This bomb alerts you to pay attention so you don't get burned.

Where to Go from Here

You can start anywhere you want. Worried about the fate of the universe? Start off with the Big Bang (see Chapter 16 if you're really interested).

Or you may want to begin with what's in store for you as you pursue your passion for the stars.

Wherever you start, I hope you continue your cosmic exploration and experience the joy, excitement, and enchantment that people have always found in the skies.

Part I
Stalking the Cosmos

The 5th Wave   By Rich Tennant

In this part . . .

Objects and events in the sky have always fascinated us humans. Throughout history, our interest in astronomy has been both practical and inspired. People navigated by the stars and planted crops according to the moon's phases (and still do today). They built sites where observations may have accompanied rituals (such as Stonehenge) and kept time by the motions of the sun and stars. And folks still wonder about the nature of objects in the heavens.

You can join this grand human tradition. In this part, I introduce the science of astronomy and offer techniques and advice for observing planets, comets, meteors, and other sights in the night sky.

Chapter 1

Seeing the Light: The Art and Science of Astronomy

In This Chapter

▶ Understanding the observational nature of astronomy

▶ Focusing on astronomy's language of light

▶ Weighing in on gravity

▶ Recognizing the movements of objects in space

Step outside on a clear night and look at the sky. If you're a city dweller or live in a cramped suburb, you see dozens, maybe hundreds, of twinkling stars. Depending on the time of the month, you may also see a full moon and up to five of the nine planets that revolve around the sun.

A shooting star or "meteor" may appear overhead. What you actually see is the flash of light from a tiny piece of comet dust streaking through the upper atmosphere.

Another pinpoint of light moves slowly and steadily across the sky. Is it a space satellite, such as the Hubble Space Telescope, or just a high-altitude airliner? If you have a pair of binoculars, you may be able to see the difference. Most airliners have running lights, and their shapes may be perceptible.

If you live out in the country — on the seashore away from resorts and developments, on the plains, or in the mountains far from any floodlit ski slope — you can see thousands of stars. The Milky Way appears as a beautiful pearly swath across the heavens. What you're seeing is the cumulative glow from millions of faint stars, individually indistinguishable with the naked eye. At a great observation place, such as Cerro Tololo in the Chilean Andes, you can see even more stars. They hang like brilliant lamps in a coal black sky, often not even twinkling, like in Van Gogh's *Starry Night* painting.

When you look at the sky, you practice astronomy — you observe the universe that surrounds you and try to make sense of what you see. For thousands of years, everything people knew about the heavens they deduced by simply observing the sky. Almost everything that astronomy deals with

- You see from a distance
- You discover by studying the light that comes to you from objects in space
- Moves through space under the influence of gravity

This chapter introduces you to these concepts (and more).

Astronomy: The Science of Observation

Astronomy is the study of the sky, the science of cosmic objects and celestial happenings, and the investigation of the nature of the universe you live in. Professional astronomers carry out the business of astronomy by observing with telescopes that capture visible light from the stars or by tuning in to radio waves that come from space. They use backyard telescopes, huge observatory instruments, and satellites that orbit Earth collecting forms of light (such as ultraviolet radiation) that the atmosphere blocks from reaching the ground. They send telescopes up in sounding rockets (equipped with instruments for making high-altitude scientific observations) and on unmanned balloons. And they send some instruments into the solar system aboard deep space probes.

Professional astronomers study the sun and the solar system, the Milky Way, and the universe beyond. They teach in universities, design satellites in government labs, and operate planetariums. They also write books (like me, your loyal *For Dummies* hero). Most have completed years of schooling to hold PhDs. So many of them study complex physics or work with automated, robotic telescopes that they have moved far beyond the night sky recognizable to our eyes. They may not have ever studied the *constellations* (groups of stars, such as Ursa Major and the Great Bear, named by ancient stargazers) that amateur or hobbyist astronomers first discover. (You may already be familiar with the Big Dipper, an *asterism* in Ursa Major. An asterism is a named star pattern outside of the 88 recognized constellations. Figure 1-1 shows the Big Dipper in the night sky.)

In addition to the more than 13,000 professional astronomers worldwide, thousands of amateur astronomers enjoy studying the skies, including more than 300,000 in the United States alone. Amateur astronomers usually know the constellations and use them as guideposts when exploring the sky by eye, with binoculars, and with telescopes. Many amateurs also make useful scientific contributions. They monitor the changing brightnesses of variable stars;

discover asteroids, comets, and exploding stars; crisscross the earth to catch the shadows cast as asteroids pass in front of bright stars (thereby helping astronomers map the asteroids' shapes); and join the search for planets orbiting stars beyond the sun.

Figure 1-1:
The Big Dipper, found in Ursa Major, is an asterism.

In the rest of Part I, I provide you with information on how to observe the skies effectively and enjoyably.

Understanding What You See: The Language of Light

Light brings us information about the planets, moons, and comets in our solar system; the stars, star clusters, and nebulae in our galaxy; and the objects beyond.

In ancient times, folks didn't think about the physics and chemistry of the stars; they absorbed and passed down folk tales and myths: the Great Bear, the Demon star, the Man in the Moon, the dragon eating the sun during a solar eclipse, and more. The tales varied from culture to culture. But many people did discover the patterns of the stars. In Polynesia, skilled navigators rowed across hundreds of miles of open ocean with no landmarks in view and no compass. They sailed by the stars, the sun, and their knowledge of prevailing winds and currents.

Gazing at the light from a star, the ancients noted its brightness, position in the sky, and color. This information helps people distinguish one sky object from another, and the ancients (and now people today) got to know them like old friends. Some basics of recognizing and describing what you see in the sky are

✔ Distinguishing stars from planets

✔ Identifying constellations, individual stars, and other sky objects by name

✔ Observing brightness (given as magnitudes)

✔ Understanding the concept of a light-year

✔ Charting sky position (measured in special units called *RA* and *Dec*)

They wondered as they wandered: Planets versus stars

The term planet comes from the ancient Greek word *planetes,* meaning wanderer. The Greeks (and other ancient people) noticed that five spots of light moved across the pattern of stars in the sky. Some moved steadily ahead; others occasionally looped back on their own paths. Nobody knew why. And these spots of light didn't twinkle like the stars did — no one understood that difference either. Every culture had a name for those five spots of light, what we now call planets. Their English names are Mercury, Venus, Mars, Jupiter, and Saturn. These celestial bodies aren't wandering through the stars; they orbit around the sun, our solar system's central star.

Today, astronomers know that planets can be smaller or bigger than Earth, but they all are much smaller than the sun. The planets in our solar system are all so close to Earth that they have perceptible disks — at least when viewed through a telescope — so we can see their shapes and sizes. The stars are so far away from Earth that even if you view them through a powerful telescope, they show up only as points of light. (For more about the planets in the solar system, flip to Part II.)

If you see a Great Bear, start worrying: Naming stars and constellations

I used to tell planetarium audiences who craned their necks to look at stars projected above them, "If you can't see a Great Bear up there, don't worry. Maybe those who *do* see a Great Bear should worry."

Ancient astronomers divided the sky into imaginary figures, such as Ursa Major (Latin for Great Bear); Cygnus, the Swan; Andromeda, the Chained Lady; and Perseus, the Hero. The ancients identified each figure with a pattern of stars. The truth is, to most people, Andromeda doesn't look much like a chained lady at all, or anything else for that matter (see Figure 1-2).

Today, astronomers have divided the sky into 88 constellations, which contain all the stars that you can see. The International Astronomical Union, which governs the science, set boundaries for the constellations so astronomers can agree on which star is in which constellation. Previously, sky maps drawn by different astronomers often disagreed. Now when you read that the Tarantula Nebula is in Dorado (see Chapter 12), you know that to see this nebula, you must seek it in the Southern Hemisphere constellation Dorado, the Goldfish.

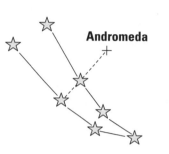

Figure 1-2:
Andromeda is also known as the Chained Lady.

Andromeda

The largest constellation is Hydra, the Water Snake. The smallest is Crux, the Cross, which most people call the "Southern Cross." You can see a Northern Cross, too, but you can't find it in a list of constellations; it's an *asterism* (a named star pattern) within Cygnus, the Swan. Although astronomers generally agree on the names of the constellations, they don't have a consensus on what each name means. For example, some astronomers call Dorado the Swordfish, but I'd like to spike that name. One constellation, Serpens, the Serpent, is broken into two sections, which aren't connected. The two sections, located on either side of Ophiuchus, the Serpent Bearer, are Serpens Caput (the Serpent's Head) and Serpens Cauda (the Serpent's Tail).

The individual stars in a constellation often have no relation to each other except for their proximity in the sky as visible from Earth. In space, the stars that make up a constellation may be completely unrelated to one another, with some located relatively near to Earth and others located at much greater distances in space. But they make a simple pattern for observers on Earth to enjoy.

As a rule, the brighter stars in a constellation were assigned a Greek letter, either by the ancient Greeks or by astronomers of later civilizations. In each constellation, the brightest star was labeled alpha, the first letter of the Greek alphabet. The next brightest star was beta, the second Greek letter, and so on down to omega, the final Greek letter of the 24-character alphabet. (The astronomers used only lowercase Greek letters, so you see them written as α, β, . . . ω.)

So Sirius, the brightest star in the night sky — in Canis Major, the Great Dog — is called Alpha Canis Majoris. (Astronomers add a suffix here or there to put star names in the Latin genitive case — scientists have always liked Latin.)

Table 1-1 shows a list of the Greek alphabet, in order, with the names of the letters and their corresponding symbols.

Table 1-1	The Greek Alphabet
Letter	*Name*
α	Alpha
β	Beta
γ	Gamma
δ	Delta
ε	Epsilon
ζ	Zeta
η	Eta
θ	Theta
ι	Iota
κ	Kappa
λ	Lambda
μ	Mu
ν	Nu
ξ	Xi
o	Omicron
π	Pi
ρ	Rho
σ	Sigma
τ	Tau
υ	Upsilon
ϕ	Phi
χ	Chi
ψ	Psi
ω	Omega

When you look at a star atlas, you discover that the individual stars in a constellation aren't marked α Canis Majoris, β Canis Majoris, and so on. Usually, the creator of the atlas marks the area of the whole constellation "Canis Major" and labels the individual stars α, β, and so on. When you read about a star in a list of objects to observe, say in an astronomy magazine (see Chapter 2), you probably won't see it listed in the style of Alpha Canis Majoris or even α Canis Majoris. Instead, to save space, the magazine prints it as α CMa; *CMa* is the three-letter abbreviation for Canis Majoris (and also the abbreviation for Canis Major). I give the abbreviation for each of the constellations in Table 1-2.

Astronomers didn't coin special names such as Sirius for every star in Canis Major, so they named them with Greek letters or other symbols. In fact, some constellations don't have a single named star. (Don't fall for those advertisements that offer to name a star for a fee. The International Astronomical Union doesn't recognize purchased star names.) In other constellations, astronomers assigned Greek letters, but they could see more than 24 stars for 24 Greek letters. Therefore, astronomers gave some stars numbers and letters from the Roman alphabet, such as 236 Cygni, b Vulpeculae, HR 1516, and worse. You may even run across RU Lupi and SX Sex. (I'm not making this up.) But like any other star, you can recognize them by their positions in the sky (as tabulated in star lists), their brightnesses, their colors, or other properties, if not their names.

When you look at the constellations today, you see many exceptions to the rule that the Greek-letter star names correspond to the respective brightnesses of the stars in a constellation. The exceptions exist because

- ✔ The letters were based on inaccurate naked-eye observations of brightness.

- ✔ Over the years, star-atlas authors changed constellation boundaries, moving some stars from one constellation into another that includes previously named stars.

- ✔ Some astronomers mapped out small and Southern Hemisphere constellations long after the Greek period, and the practice wasn't always followed.

- ✔ The brightness of some stars has changed over the centuries since the ancient Greeks charted them.

A good (or bad) example is the constellation Vulpecula, the Fox, where only one of the stars (alpha) has a Greek letter.

Because alpha isn't always the brightest star in a constellation, astronomers needed another term to describe that exalted status, and *lucida* is the word (from the Latin word *lucidus,* meaning bright or shining). The lucida of Canis Major is Sirius, the alpha star, but the lucida of Orion, the Hunter, is Rigel, which is Beta Orionis. The lucida of Leo Minor, the Little Lion (a particularly inconspicuous constellation), is 46 Leo Minoris.

Table 1-2 lists the 88 constellations, the brightest star in each, and the magnitude of that star. *Magnitude* is a measure of a star's brightness. (I talk about magnitudes later in this chapter in the section "The smaller, the brighter: Getting to the root of magnitudes.") When the lucida of a constellation is the alpha star and has a name, I list only the name. For example, in Auriga, the Charioteer, the brightest star, Alpha Aurigae, is Capella. But when the lucida isn't an alpha, I give its Greek letter or other designation in parentheses. For example, the lucida of Cancer, the Crab, is Al Tarf, which is Beta Cancri.

Table 1-2		The Constellations and Their Brightest Stars		
Name	*Abbreviation*	*Meaning*	*Star*	*Magnitude*
Andromeda	And	Chained Lady	Alpheratz	2.1
Antlia	Ant	Air Pump	Alpha Antliae	4.3
Apus	Aps	Bird of Paradise	Alpha Apodis	3.8
Aquarius	Aqr	Water Bearer	Sadalmelik	3.0
Aquila	Aql	Eagle	Altair	0.8
Ara	Ara	Altar	Beta Arae	2.9
Aries	Ari	Ram	Hamal	2.0
Auriga	Aur	Charioteer	Capella	0.1
Bootes	Boo	Herdsman	Arcturus	−0.04
Caelum	Cae	Chisel	Alpha Caeli	4.5
Camelopardalis	Cam	Giraffe	Beta Camelopardalis	4.0
Cancer	Cnc	Crab	Al Tarf (Beta Cancri)	3.5
Canes Venatici	CVn	Hunting Dogs	Cor Caroli	2.8
Canis Major	CMa	Great Dog	Sirius	−1.5
Canis Minor	CMi	Little Dog	Procyon	0.4
Capricornus	Cap	Goat	Deneb Algedi (Delta Capricorni)	2.9
Carina	Car	Ship's Keel	Canopus	−0.7
Cassiopeia	Cas	Queen	Schedar	2.2
Centaurus	Cen	Centaur	Rigil Kentaurus	−0.3
Cepheus	Cep	King	Alderamin	2.4

Name	Abbreviation	Meaning	Star	Magnitude
Cetus	Cet	Whale	Deneb Kaitos (Beta Ceti)	2.0
Chamaeleon	Cha	Chamaeleon	Alpha Chamaeleontis	4.1
Circinus	Cir	Compasses	Alpha Circini	3.2
Columba	Col	Dove	Phakt	2.6
Coma Berenices	Com	Berenice's Hair	Beta Comae Berenices	4.3
Corona Australis	CrA	Southern Crown	Alpha Coronae Australis	4.1
Corona Borealis	CrB	Northern Crown	Alphekka	2.2
Corvus	Crv	Crow	Gienah (Gamma Corvi)	2.6
Crater	Crt	Cup	Delta Crateris	3.6
Crux	Cru	Cross	Acrux	0.7
Cygnus	Cyg	Swan	Deneb	1.3
Delphinus	Del	Dolphin	Rotanev (Beta Delphini)	3.6
Dorado	Dor	Goldfish	Alpha Doradus	3.3
Draco	Dra	Dragon	Thuban	3.7
Equuleus	Equ	Little Horse	Kitalpha	3.9
Eridanus	Eri	River	Achernar	0.5
Fornax	For	Furnace	Alpha Fornacis	3.9
Gemini	Gem	Twins	Pollux (Beta Geminorum)	1.1
Grus	Gru	Crane	Alnair	1.7
Hercules	Her	Hercules	Ras Algethi	2.6
Horologium	Hor	Clock	Alpha Horologii	3.9
Hydra	Hya	Water Snake	Alphard	2.0
Hydrus	Hyi	Little Water Snake	Beta Hydri	2.8

(continued)

Table 1-2 *(continued)*

Name	Abbreviation	Meaning	Star	Magnitude
Indus	Ind	Indian	Alpha Indi	3.1
Lacerta	Lac	Lizard	Alpha Lacertae	3.8
Leo	Leo	Lion	Regulus	1.4
Leo Minor	LMi	Little Lion	Praecipua (46 Leo Minoris)	3.8
Lepus	Lep	Hare	Arneb	2.6
Libra	Lib	Scales	Zubeneschemali (Beta Librae)	2.6
Lupus	Lup	Wolf	Alpha Lupus	2.3
Lynx	Lyn	Lynx	Alpha Lyncis	3.1
Lyra	Lyr	Lyre	Vega	0.0
Mensa	Men	Table	Alpha Mensae	5.1
Microscopium	Mic	Microscope	Gamma Microscopii	4.7
Monoceros	Mon	Unicorn	Beta Monocerotis	3.7
Musca	Mus	Fly	Alpha Muscae	2.7
Norma	Nor	Level and Square	Gamma Normae	4.0
Octans	Oct	Octant	Nu Octantis	3.8
Ophiuchus	Oph	Serpent Bearer	Rasalhague	2.1
Orion	Ori	Hunter	Rigel (Beta Orionis)	0.1
Pavo	Pav	Peacock	Alpha Pavonis	1.9
Pegasus	Peg	Winged Horse	Enif (Epsilon Pegasi)	2.4
Perseus	Per	Hero	Mirphak	1.8
Phoenix	Phe	Phoenix	Ankaa	2.4
Pictor	Pic	Easel	Alpha Pictoris	3.2
Pisces	Psc	Fish	Eta Piscium	3.6
Pisces Austrinus	PsA	Southern Fish	Fomalhaut	1.2

Name	Abbreviation	Meaning	Star	Magnitude
Puppis	Pup	Ship's Stern	Zeta Puppis	2.3
Pyxis	Pyx	Compass	Alpha Pyxidus	3.7
Reticulum	Ret	Net	Alpha Reticuli	3.4
Sagitta	Sge	Arrow	Gamma Sagittae	3.5
Sagittarius	Sgr	Archer	Kaus Australis (Epsilon Sagittarii)	1.9
Scorpius	Sco	Scorpion	Antares	1.0
Sculptor	Scl	Sculptor	Alpha Sculptoris	4.3
Scutum	Sct	Shield	Alpha Scuti	3.9
Serpens	Ser	Serpent	Unukalhai	2.7
Sextans	Sex	Sextant	Alpha Sextantis	4.5
Taurus	Tau	Bull	Aldebaran	0.9
Telescopium	Tel	Telescope	Alpha Telescopium	3.5
Triangulum	Tri	Triangle	Beta Trianguli	3.0
Triangulum Australe	TrA	Southern Triangle	Alpha Trianguli Australis	1.9
Tucana	Tuc	Toucan	Alpha Tucanae	2.9
Ursa Major	UMa	Great Bear	Alioth (Epsilon Ursae Majoris)	1.8
Ursa Minor	UMi	Little Bear	Polaris	2.0
Vela	Vel	Sails	Suhail al Muhlif (Gamma Velorum)	1.8
Virgo	Vir	Virgin	Spica	1.0
Volans	Vol	Flying Fish	Gamma Volantis	3.6
Vulpecula	Vul	Fox	Anser	4.4

Identifying stars would be much easier if they had little name tags that you could see through your telescope, but at least they don't have unlisted numbers like an old friend that you're desperate to reach. (For the whole scoop on stars, check out Part III.)

What do I spy? The Messier Catalog and other sky objects

Naming stars was easy enough for astronomers. But what about all those other objects in the sky — galaxies, nebulae, star clusters, and the like (which I cover in Part III). Charles Messier, a French astronomer in the late 18th century, created a list of about 100 fuzzy sky objects and gave them numbers. His list is known as the *Messier Catalog,* and now when you hear the Andromeda Galaxy called by its scientific name, M31, you know what it means. Today, 110 objects make up the standard Messier Catalog.

You can find pictures and a complete list of the Messier objects at The Messier Catalog Web site of Students for the Exploration and Development of Space at `www.seds.org/messier`. And you can find out how to earn a certificate for viewing Messier objects from the Astronomical League Messier Club Web site at `www.astroleague.org/al/obsclubs/messier/mess.html`.

Experienced amateur astronomers often engage in Messier marathons in which each person tries to observe every object in the *Messier Catalog* during a single long night. But in a marathon, you don't have time to enjoy an individual nebula, star cluster, or galaxy. My advice is to take it slow and savor their individual visual delights. A wonderful book on the Messier objects, which includes hints on how to observe each object, is Stephen J. O'Meara's *The Messier Objects* (Cambridge University Press and Sky Publishing Corporation, 1998).

Astronomers have confirmed the existence of thousands of other *deep sky objects,* the term amateurs use for star clusters, nebulae, and galaxies to distinguish them from stars and planets. Because Messier didn't list them, astronomers refer to these objects by their numbers as given in catalogues made since his time. You can find many of these objects listed in viewing guides and sky maps by their NGC *(New General Catalogue)* and IC *(Index Catalogue)* numbers. For example, the bright double cluster in Perseus, the Hero, consists of NGC 869 and NGC 884.

The smaller, the brighter: Getting to the root of magnitudes

A star map, constellation drawing, or list of stars always indicates each star's magnitude. The *magnitudes* represent the brightnesses of the stars. One of the ancient Greeks, Hipparchos (also spelled Hipparchus, but he wrote it in Greek), divided all the stars he could see into six classes. He called the brightest stars magnitude one or *1st magnitude,* the next brightest bunch the *2nd magnitude* stars, and on down to the dimmest ones, which were *6th magnitude.*

TECHNICAL STUFF

By the numbers: The mathematics of brightness

The 1st magnitude stars are about 100 times brighter than the 6th magnitude stars. In particular, the 1st magnitude stars are about 2.512 times brighter than the 2nd magnitude stars, which are about 2.512 times brighter than the 3rd magnitude stars, and so on. (At the 6th magnitude, you get up into some big numbers: 1st magnitude stars are about 100 times brighter.) You mathematicians out there recognize this as a *systematic progression*. Each magnitude is the 5th root of 100 (meaning that when you multiply a number by itself four times — for example, 2.512 $\times$ 2.512 $\times$ 2.512 $\times$ 2.512 $\times$ 2.512 — the result is 100). If you doubt my word and do this calculation on your own, you get a slightly different answer because I left off some decimal places.

Thus, you can calculate how faint a star is — compared to some other star — from its magnitude. If two stars are 5 magnitudes apart (such as the 1st magnitude star and the 6th magnitude star), they differ by a factor of 2.512^5 (2.512 to the fifth power), and a good pocket calculator shows you that one star is 100 times brighter. If two stars are 6 magnitudes apart, one is about 250 times brighter than the other. And if you want to compare, say, a 1st magnitude star with an 11th magnitude star, you factor a 2.512^{10} difference in brightness, meaning a factor of 100 squared, or 10,000.

The faintest object visible with the Hubble Space Telescope is about 25 magnitudes fainter than the faintest star you can see with the naked eye (assuming normal vision and viewing skills — some experts and a certain number of liars and braggarts say that they can see 7th magnitude stars). Speaking of dim stars, 25th magnitudes are five times 5 magnitudes, which corresponds to a brightness difference of a factor of 100^5. So the Hubble can see $100 \times 100 \times 100 \times 100 \times 100$, or 10 billion times fainter than the human eye. Astronomers expect nothing less from a billion-dollar telescope. At least it didn't cost $10 billion.

You can get a good telescope for under $1,000, and you can download the billion-dollar Hubble's best photos from the Internet for free at www.stsci.edu.

Notice that contrary to most common measurement scales and units, the brighter the star, the smaller the magnitude. The Greeks weren't perfect, however; even Hipparchos had an Achilles' heel: He didn't leave room in his system for the very brightest stars, when accurately measured.

So, today, we recognize a few stars with a zero magnitude or a negative magnitude. Sirius, for example, is magnitude –1.5. And the brightest planet, Venus, is sometimes magnitude –4 (the exact value differs, depending on the distance Venus is from Earth at the time and its direction with respect to the sun).

Another omission: Hipparchos didn't have a magnitude class for stars that he couldn't see. This didn't seem like an oversight at the time, because nobody knew about these stars. But today, astronomers know that millions of stars exist beyond our naked-eye view, and they all should have magnitudes. Their magnitudes are larger numbers: 7 or 8 for stars easily seen through binoculars and 10 or 11 for stars easily seen through a good, small telescope. The magnitudes reach as high (and as dim) as 21 for the faintest stars in the Palomar Observatory Sky Survey and 30 or maybe 31 for the faintest objects imaged with the Hubble Space Telescope.

Looking back on light-years

The distances to the stars and other objects beyond the planets of our solar system are measured in *light-years*. As a measurement of actual length, a light-year is about 5.9 trillion miles long.

People confuse a light-year with a length of time, because the term contains the word "year." But a light-year is really a distance measurement — the length that light travels, zipping through space at 186,000 miles per second, over the course of a year.

When you view an object in space, you see it as it appeared when the light left the object. Consider these examples:

- ✔ When astronomers spot an explosion on the sun, we don't see it in real time; the light from the explosion takes about eight minutes to get to Earth.

- ✔ The nearest star beyond the sun, Proxima Centauri, is about 4 light-years away. Astronomers can't see Proxima as it is now, only as it was four years ago.

- ✔ Look up at the Andromeda Galaxy, the most distant object that you can readily see with the unaided eye, on a clear, dark night in the fall. The light your eye receives left that galaxy about 2.6 million years ago. If the galaxy disappeared by some mysterious means, we wouldn't even know for over two million years. (See Chapter 12 for more hints on viewing galaxies.)

Here's the bottom line:

- ✔ When you look out into space, you're looking back in time.

- ✔ Astronomers don't have a way to know exactly what an object out in space looks like right now.

When you look at some big, bright stars in a faraway galaxy, you must entertain the possibility that those particular stars don't even exist any more. Some massive stars only live for 10 or 20 million years. If you see them in a galaxy that exists 50 million light-years away, you're looking at lame duck stars. They aren't shining in that galaxy any more; they're dead.

If astronomers send a flash of light toward one of the most distant galaxies found with Hubble and other major telescopes, the light would take at least 12 billion years to arrive, because the farthest galaxies are at least 12 billion light-years away (furthermore, the universe is expanding, so those galaxies will be farther away by the time the light gets there). Astronomers, however, project that the sun will swell up and destroy all life on Earth a mere five or six billion years from now, so the light would be a futile advertisement of our civilization's existence, a flash in the celestial pan.

Keep on moving: Figuring the positions of the stars

Astronomers used to call stars "fixed stars" to distinguish them from the wandering planets. But, in fact, stars are in constant motion as well, both real and apparent. The whole sky rotates overhead because the earth is turning. The stars rise and the stars set, like the sun and the moon, but they stay in formation. The stars that make up the Great Bear don't swing over to the Little Dog or Aquarius, the Water Bearer. Different constellations rise at different times and on different dates as visible from different places around the globe.

Actually, the stars in Ursa Major (and every other constellation) do move with respect to one another and at breathtaking speeds, measured in hundreds of miles per second. But those stars are so far away that scientists need precise measurements over considerable intervals of time to detect their motions across the sky. So, 20,000 years from now, the stars in Ursa Major will form a different pattern in the sky. Maybe they will even look like a Great Bear.

Hey, you! No, no, I mean A.U.

The Earth is about 93 million miles from the sun, or one *Astronomical Unit* (A.U.). The distances between objects in the solar system are usually given in A.U. Its plural is also A.U. (Don't confuse A.U. with "Hey, you!")

In public announcements, press releases, and popular books, astronomers state how far the stars and galaxies that they study are "from Earth." But among themselves and in technical journals, they always give the distances from the sun, the center of our solar system. This rarely matters, because astronomers can't measure the distances of the stars precisely enough for one A.U. more or less to make a difference, but they do it this way for consistency.

In the meantime, astronomers have measured the positions of millions of stars, and many of them are tabulated in catalogs and marked on star maps. The positions are listed in a system called "right ascension and declination" — known to all astronomers, amateur and pro, as *RA* and *Dec:*

- The RA is the position of a star measured in the east-west direction on the sky (like longitude, the position of a place on Earth measured east or west of the prime meridian at Greenwich, England).

- The Dec is the position of the star measured in the north-south direction, like the latitude of a city, which is measured north or south of the equator.

Astronomers usually list RA in units of hours, minutes, and seconds, like time. We list Dec in degrees, minutes, and seconds of arc. Ninety degrees make up a right angle, sixty minutes of arc make up a degree, and sixty seconds of arc create a minute of arc. A minute or second of arc is also often called an "arc minute" or an "arc second," respectively.

TECHNICAL STUFF

Digging deeper into RA and Dec

A star at RA 2h00m00s is two hours east of a star at RA 0h00m00s, regardless of their declinations. RA increases from west to east, starting from RA 0h00m00s, which corresponds to a line in the sky (actually half a circle, centered on the center of the earth) from the North Celestial Pole to the South Celestial Pole. The first star may be at Dec 30° North and the second star may be at Dec 15° 25'12" South, but they're still two hours apart in the east-west direction (and 45° 25'12" apart in the north-south direction). The North and South Celestial Poles are the points in the sky — due north and due south — around which the whole sky seems to turn, with the stars all rising and setting.

See the following details about the units of RA and Dec:

- An hour of RA equals an arc of 15 degrees on the equator in the sky. Twenty-four hours of RA span the sky, and 24 × 15 = 360 degrees, or a complete circle around the sky. A minute of RA, called a *minute of time,* is a measure of angle on the sky that makes up ⅟₆₀ of an hour of RA. So you take 15° ÷ 60, or ¼°. A second of RA, or a *second of time,* is 60 times smaller than a minute of time.

- Dec is measured in degrees, like the degrees in a circle, and in minutes and seconds of arc. A whole degree is about twice the apparent or angular size of the full Moon. Each degree is divided into 60 seconds of arc. The sun and the full Moon each are about 32 minutes of arc (32') wide as seen on the sky, although in reality the sun is much larger than the moon. Each minute of arc is divided into 60 seconds of arc (60"). When you look through a backyard telescope at high magnification, turbulence in the air blurs the image of a star. Under good conditions (low turbulence), the image should measure about 1" or 2" across. That's one or two arc seconds, not one or two inches.

A few simple rules may help you remember how RA and Dec work and how to read a star map (see Figure 1-3):

- The North Celestial Pole (NCP) is the place to which the axis of the earth points in the north direction. If you stand at the geographic North Pole, the NCP is right overhead. (If you stand there, say "Hi" to Santa for me, but beware: You may be standing on thin ice; no land covers the geographic North Pole.)

- The South Celestial Pole (SCP) is the place to which the axis of the earth points in the south direction. If you stand at the geographic South Pole, the SCP is right overhead. I hope you dressed warmly: You're in Antarctica!

- The imaginary lines of equal RA run through the NCP and SCP as semicircles centered on the center of Earth. They may be imaginary, but they appear marked on most sky maps to help people find the stars at particular RAs.

- The imaginary lines of equal Dec, like the line in the sky that marks Dec 30° North, pass overhead at the corresponding geographic latitudes. So if you stand in New York City, latitude 41° North, the point overhead is always at Dec 41° North, although its RA changes constantly as the earth turns. These imaginary lines appear on star maps, too, as *declination circles*.

The Celestial Sphere

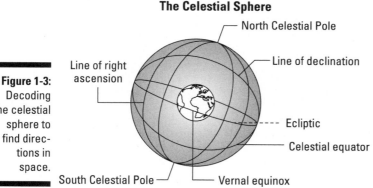

Figure 1-3: Decoding the celestial sphere to find directions in space.

Suppose you want to find the NCP as visible from your backyard. Face due north and look at an altitude of *x* degrees, where *x* is your geographic latitude. I'm assuming that you live in North America, Europe, or somewhere in the Northern Hemisphere. If you live in the Southern Hemisphere, you can't see the NCP. You can, however, look for the SCP. Look for the spot due south whose altitude in the sky, measured in degrees above the horizon, is equal to your geographic latitude.

In almost every astronomy book, the symbol " means seconds of arc, not inches. But at every university, a student in Astronomy 101 always writes on an exam, "The image of the star was about one inch in diameter." Understanding beats memorizing every day, but not everyone understands.

Here's the good news: If you just want to spot the constellations and the planets, you don't have to know how to use RA and Dec. You just have to compare a star map prepared for the right time of year and time of night (as printed, for example, in many daily newspapers) with the stars that you see at the corresponding times. But if you want to understand how star catalogs and maps work and how to zero in on faint galaxies with your telescope, understanding the system helps.

And if you purchase one of those snazzy, new, surprisingly affordable telescopes with computer control (see Chapter 3), you can punch in the RA and Dec of a recently discovered comet and the scope points right at it. (A little table called an *ephemeris* comes with every announcement of a new comet. It gives the predicted RA and Dec of the comet on successive nights as it sweeps across the sky.)

Gravity: A Force to Be Reckoned With

Ever since the work of Newton (Isaac, not the cookie), everything in astronomy has revolved around gravity. Newton explained it as a force between any two objects. The force depends on mass and separation. The more massive the object, the more powerful its pull. The greater the distance, the weaker the gravitational attraction.

Albert Einstein developed an improved theory of gravity, which passes experimental tests that Isaac's theory flunks. Newton's theory was good enough for commonly experienced gravity, like the force that made the apple fall on his head (if it really did hit him). But Einstein's theory also predicts effects that happen close to massive objects, where gravity is very strong. Einstein didn't think of gravity as a force; he considered it the bending of space and time by the very presence of a massive object, such as a star. I get all bent out of shape just thinking about it.

Newton's concept of gravity explains the following:

- ✔ Why the moon orbits the earth, why the earth orbits the sun, why the sun orbits the center of the Milky Way, and why many other objects orbit one object or another out there in space

- ✔ Why a star or a planet is round

- ✔ Why gas and dust in space may clump together to form new stars

Einstein's theory of gravity, the General Theory of Relativity, explains the following:

- Why stars visible near the sun during a total eclipse seem slightly out of position

- Why black holes exist

- Why, as the earth turns, it drags warped space and time around with it, an effect that scientists have verified with the help of satellites orbiting the earth.

You can find out about black holes in Chapters 11 and 13 without mastering the General Theory of Relativity. You can become smarter if you read every chapter in this book, but your friends won't call you Einstein unless you let your hair grow, parade around in a messy old sweater, and stick out your tongue when they take your picture.

Space: A Commotion of Motion

Everything in space is moving and turning. Objects can't sit still. Thanks to gravity, a celestial body is always pulling on a star, planet, galaxy, or spacecraft. The universe has no center.

For example, the earth

- Turns on its axis — what astronomers call *rotating* — and takes one day to turn all the way around

- Orbits around the sun — what astronomers call *revolving* — with one complete orbit taking one year

- Travels with the sun in a huge orbit around the center of the Milky Way; the trip takes about 226 million years to complete once, and the duration of the trip is called the *galactic year*

- Moves with the Milky Way in a trajectory around the center of the *Local Group of Galaxies,* a couple dozen galaxies in our neck of the universe

- Moves through the universe with the Local Group as part of the *Hubble Flow,* the general expansion of space caused by the Big Bang

  The Big Bang is the event that gave rise to the universe and set space itself expanding at a furious rate. Detailed theories about the Big Bang explain many observed phenomena and have successfully predicted some that hadn't been observed before the theories were circulated. (For more about the Big Bang and other aspects of the universe, check out Part IV.)

Remember Ginger Rogers? She did everything Fred Astaire did when they danced in the movies, and she did it all backward. Like Ginger and Fred, the moon follows all the motions of the earth (although not backward), except for the earth's rotation; the moon rotates more slowly, about once a month. And it performs its tasks while also revolving around the earth (which it does about once a month).

And you, as a person on Earth, participate in the motions of rotation, revolution, galactic orbiting, Local Group cruising, and cosmic expansion. You do all that while you drive to work without even knowing it. Ask your boss for a little consideration the next time you run a few minutes late.

Chapter 2

Join the Crowd: Skywatching Activities and Resources

In This Chapter

▶ Joining astronomy clubs, using the Internet, and more

▶ Exploring observatories and planetariums

▶ Enjoying star parties, eclipse tours and cruises, and telescope motels

*A*stronomy has universal appeal. The stars have fascinated people everywhere from prehistoric times into the modern age. Early observations of the sky led to all sorts of theories about the universe and attributions of power and purpose to the movements of stars, planets, and comets. As you look up at the sky, hundreds of thousands of people worldwide watch with you. When it comes to skywatching, you're not alone. Many people, publications, and other resources are at your disposal to help you get started, to keep you going, and to help you participate in the great work of explaining the universe.

In this chapter, I introduce you to these resources and give you suggestions for how you can get started. The rest is up to you. So join in!

After you know the resources, organizations, facilities, and equipment that can help you enjoy astronomy more deeply, you can move comfortably on to the science of astronomy itself — the nature of the objects and phenomena out there in deep space. I describe the equipment you need in order to get started in Chapter 3.

You're Not Alone: Astronomy Clubs, Web Sites, and More

You have plenty of readily available information, organizations, people, and facilities to help you get started and remain active in astronomy. You can join associations and activities to help researchers keep track of stars and planets, and you can attend astronomy club meetings, lectures, and instructional sessions, which allow you to share telescopes and viewing sites to enjoy the sky with others. You also can find Web sites, publications, and computer programs with basic information on astronomy and current events in the sky.

Joining an astronomy club for star-studded company

The best way to break into astronomy without undue effort and expense is to join an astronomy club and meet the regulars. Clubs hold monthly meetings where the old hands pass on tips on techniques and equipment to beginners and where local and visiting scientists present talks and slide shows. Members should already know where to get a good deal on a used telescope or binoculars and which products on the market are worth the money. (See Chapter 3 for more.)

Even better, astronomy clubs sponsor observing meetings, usually on weekend nights and occasionally on special dates, when a meteor shower, an eclipse, or another special event occurs. An observing meeting is the best place to find out about the practice of astronomy and the equipment you need. You don't even need to bring a telescope; most folks are happy to give you a look through theirs. Just wear sensible shoes, bring mittens and a hat for the cool night air, and put on a smile!

If you live in a city or a suburb, chances are good that your night sky is bright, so you can find better observational conditions if you travel to a dark spot in the country. Your local astronomy club probably already has a good spot, and when the members converge on that lonely place, you can enjoy safety in numbers.

If you live in a good-sized city or a college town, an astronomy club should be nearby. To find it, just hop on the Web and use the locator form at skyand telescope.com (just put your mouse on the "Resources" tab to find clubs and organizations). Insert your city, state or province, and country, and

Gazing around the world: A sampling of astronomy clubs

The Astronomical Society of the Pacific (www.astrosociety.org), with headquarters in San Francisco, publishes the magazine *Mercury* for amateurs. It holds an annual meeting that moves around the western United States and sometimes goes as far east as Boston or Toronto. It offers numerous educational materials in astronomy to schoolteachers as well.

Do you live in Canada? The Royal Astronomical Society of Canada has 27 *Centres,* which is a fancy name for astronomy clubs. Often, professionals from the nearest university get involved with the Centre's activities. You can find a list of Centres on the RASC Web site, www.rasc.ca.

In the United Kingdom, the organization of choice is the British Astronomical Association, founded in 1890 and still going strong. Its Web site is www.britastro.org/main/.

Most other countries have astronomy clubs, too. Astronomy is truly a "universal" passion.

information on the nearest amateur astronomy group(s) pops up. In Kansas City, Missouri, for example, skywatcher wannabes have two groups to choose from, and in Vancouver, British Columbia, you can join a club with over 300 members.

You can also check out the Web site of America's "club of clubs," the Astronomical League, at www.astroleague.org. Browse through the list of over 240 member societies, arranged by state.

Checking Web sites, magazines, and software

Finding out about astronomy is easy. You can choose from a wide range of resources, including Web sites, magazines, and some innovative new software. The following sections offer some tips for finding the best information.

Traveling through cyberspace

The Net offers sites on every topic in astronomy, and the resources are increasing at, well, an astronomical rate! You can find many Web sites listed throughout this book; if you want more information on planets, comets, meteors, or eclipses, the Web offers good sites on every topic.

The editors of *Sky & Telescope* magazine maintain one of the best Web sites at skyandtelescope.com. Get your observational career started by checking out Sky & Telescope's "This Week's Sky at a Glance" page at skyandtelescope.com/observing/ataglance/article_110_1.asp. It gives a day by day (or night by night) account of planets, comets, and other sights to look for.

Astronomy leaving you completely puzzled? At some Web sites, NASA scientists or other researchers are on hand to answer your questions. For questions about the sun and how it affects the earth, try "Ask the Space Scientist" at image.gsfc.nasa.gov/poetry/ask/askmag.html. You can pose questions about black holes and quasars to the "High Energy Astronomer" at imagine.gsfc.nasa.gov/docs/ask_astro/ask_an_astronomer.html. A child can use the "Ask A Scientist" Web site at www.windows.ucar.edu/tour/link=/kids_space/ask.html and check back to look for a posted response to his or her question for everyone to see.

Perusing publications

You can purchase excellent magazines to expand your knowledge of astronomy and your skill at practicing it. Most amateur astronomers subscribe to at least one publication. And, in many cases, if you join a local astronomy club, a subscription to a national magazine may be available at a member's discount. (See "Joining an astronomy club for star-studded company," earlier in this chapter, for the scoop on clubs.)

Try *Night Sky,* which the creators gear toward the complete astronomy beginner. It goes to press just six times per year, so you have time to get up to speed on the information in one issue before the next arrives. Its Web site, www.nightskymag.com, tells you how to subscribe online, by telephone, or by e-mail. And you can find the publication at many planetarium and museum bookshops.

When you're ready to graduate from *Night Sky,* pick up a copy of each of the "big two" (literally, the biggest two) astronomical magazines: *Sky & Telescope* (which publishes *Night Sky*) and *Astronomy.* Test-drive the publications for a month, and if you get more out of one than the other, go ahead and subscribe. You can do so from their Web sites at skyandtelescope.com and www.astronomy.com.

Canadian readers can get the bimonthly *SkyNews: The Canadian Magazine of Astronomy & Stargazing,* a slick, full-color publication available from the National Museum of Science and Technology Corporation. Call 1-866-759-0005 or visit www.skynewsmagazine.com for information.

In France, a popular magazine is *Ciel & Espace* (www.cieletespace.fr); in Australia, *Sky & Space* (www.skyandspace.com.au) rules the roost (but as of press time, a new publication called *Sky & Telescope Australia* is being launched — by the time you read this, it may have a Web site that you can find on the Internet); and in Germany, look for *Sterne und Weltraum* (www.wissenschaft-online.de/suw). Wherever you live, you can find an astronomy magazine just for you.

And wherever you live, you need the annual *Observer's Handbook* of the Royal Astronomical Society of Canada (www.rasc.ca). Dozens of experts compile the handbook to help you enjoy the skies.

Surveying software

A planetarium program for your personal computer is a real plus. The program can show you what the sky looks like from your home every day. This software is terrific to look at before you step outside to view the night sky. Some astronomers use these programs to plan their observing sessions. They prepare schedules of objects to scan with telescopes and binoculars at different times of the night to use their "dark time" effectively.

Personal planetarium programs are available over a wide price range with many different features. You can find the current versions nationally advertised in astronomy and science magazines and on Web sites (see the previous two sections); they've been updated for increased usefulness. You need only one program to get started, and that may be the only program you ever need. The best way to select the planetarium program that suits you is to talk to experienced amateur astronomers at your local astronomy club. What works for them should work for you.

Here are a few programs that I find useful, each of which needs a CD-ROM drive on your computer:

✔ Although I've paid $300 or more for planetarium programs, the one I've recently enjoyed using retails for under $60: the cheapest and simplest version of *Starry Night,* from the folks at www.space.com.

   After you install the program, click on the *Starry Night* icon and up comes a color picture of the sky (a simulation, not a live image), complete with a horizon and a few trees.

   If you access the program at night, the picture shows how the stars, planets, and the moon look to the naked eye outdoors at your location (assuming the weather is clear). If you run the program during the day, you get a picture of the blue sky, with the sun at the proper altitude. Turn on the program shortly after sunset and you see the sky darken and the planets and stars come out minute by minute. You get a realistic view of the outside world without leaving your computer! But you won't find any red sails in this sunset; the program is scientific, not romantic. Click on icons, drag objects in the computer sky with your mouse, or change some settings and you can see what the sky looks like at any place around the world at almost any time.

✔ *TheSky* is a highly regarded planetarium program with plenty of bells and whistles. You can purchase the program in several versions, for beginners and advanced astronomers, at prices from $49 to $279. Software Bisque in Golden, Colorado produces the program, and you can check it out on the Web site www.bisque.com, which tells about the many features.

✔ You can find a very simple (and free!) online planetarium program at skyandtelescope.com. An *Interactive Sky Chart* comes up on your screen, showing the sky over Greenwich, England at the present time. You can reset it for your own geographic location and for whatever time and date you plan to go stargazing.

Visiting Observatories and Planetariums

You can visit *professional observatories* (organizations that have large telescopes staffed by astronomers and other scientists for use in studying the universe) and *public planetariums* (specially equipped facilities with machines that project stars and other sky objects in a darkened room with plain English explanations of various sky phenomena) to find out more about telescopes, astronomy, and research programs.

Ogling the observatories

You can find dozens of professional observatories in the United States and many more abroad. Some serve as research institutions operated by colleges and universities or government agencies. Examples include the U.S. Naval Observatory (located in the heart of Washington, D.C., guarded by the Secret Service, and home to the vice president of the United States; www.usno.navy.mil) and outposts on remote mountaintops (such as the University of Denver's Mt. Evans Meyer-Womble Observatory, billed as the "Highest Operating Observatory in the West" at 14,148 feet; www.du.edu/~rstencel/MtEvans). Other observatories are dedicated entirely to public education and information; cities, counties, school systems, or nonprofit organizations often operate these sites.

The research observatories are usually located in exotic or aptly named places, such as the Lowell Observatory (www.lowell.edu) on Mars Hill in Flagstaff,

Arizona, where the planet Pluto was discovered in 1930. Observatory founder Percival Lowell thought he could see canals on Mars through a telescope on the hill.

The National Solar Observatory (www.sunspot.noao.edu) runs a cluster of sunwatching telescopes at Sunspot, New Mexico, high above the little town of Cloudcroft, which is high above the city of Alamogordo. And Georgia State University (www.chara.gsu.edu/HLCO) operates an observatory about 50 miles east of Atlanta at Hard Labor Creek. (Don't trespass there!)

Some observatories are located in or near busy cities, such as the Griffith Observatory and Planetarium (www.griffithobs.org) — operated entirely for the public at Griffith Park in Los Angeles — and the Mount Wilson Observatory (www.mtwilson.edu) in the San Bernadino mountains above Los Angeles, California. Mount Wilson, where the expansion of the universe and the magnetism of the sun were discovered, is well worth a visit.

At Palomar Observatory (www.astro.caltech.edu/observatories/palomar/), near San Diego, California, you can see the famous 200-inch telescope, which for decades was the largest and best in the world. Even with newer instrumentation, the telescope is still a great contributor of new knowledge on the universe. The observatory has a small museum and gift shop, too.

One of the largest observatories in the continental United States is the Kitt Peak National Observatory (www.noao.edu/kpno/kpno.html), 56 miles west of Tucson, Arizona. The National Optical Astronomy Observatory (www.noao.edu) runs it. And if you want another large source of optical telescopes, you can find many of the largest and most advanced telescopes from the United States, Canada, Japan, and the United Kingdom — known collectively as the Mauna Kea Observatories (www.ifa.hawaii.edu/mko/maunakea.htm) — atop the Mauna Kea volcano on the island of Hawaii.

The Maria Mitchell Observatory (http://209.68.19.123/), on Nantucket Island off Cape Cod, Massachusetts, offers summer astronomy classes for children, lectures, and "open observation nights."

You can also visit radio astronomy observatories, where scientists "listen" to radio signals from the stars or even seek signals from alien civilizations. The National Radio Astronomy Observatory (www.nrao.edu), for example, has facilities near Socorro, New Mexico and Green Bank, West Virginia that welcome visitors.

You can find a list of observatories on the *Sky & Telescope* Web site by using the same form for locating an astronomy club (see the section "Joining an astronomy club for star-studded company" earlier in this chapter). Just go to skyandtelescope.com/resources/organizations, check the box labeled "Observatory," leave the box for a city name blank, and choose a country. If you click the United States, you bring up a list of over 250 observatories.

Or you can take a look at the AstroWeb directory of Observatories and Telescopes at cdsweb.u-strasbg.fr/astroweb/telescope.html. The directory covers the whole world, plus telescopes in space (don't try to visit them).

Many observatories have an open house for the public at weekly or monthly intervals, and some even offer daily tours (during the day) and operate astronomical museums, complete with souvenir shops.

Popping in on planetariums

Planetariums, also called *planetaria,* are just right for a beginning astronomer. They provide instructive exhibits and project wonderful sky shows indoors on the planetarium dome or on a huge screen. And many offer nighttime sky-watching sessions with small telescopes, usually held outside in the parking lot, in an adjacent small observatory dome, or at a nearby public park. Many have excellent shops where you can browse through the latest astronomy books, magazines, and star charts. The planetarium staff can also direct you to the nearest astronomy club, which may even meet after hours in the planetarium itself.

I practically grew up in the Hayden Planetarium at the American Museum of Natural History in New York City. Occasionally, I confess, I even snuck in for free. The planetarium staff was nice enough to have me back to speak (also for free) at their 50th anniversary. Although the old planetarium has been torn down, a spectacular new one has replaced it. You should make this planetarium a prime destination next time you visit the Big Apple. It's pricey, but still much cheaper than a Broadway show, and its stars never miss a cue or sing off key (just don't try to sneak in like me!).

Loch Ness Productions, in Massachusetts, is the monster of the planetarium business. It keeps tabs on over 2,500 planetariums worldwide. To find a planetarium near you, check out the *Planetarium Compendium* at www.lochness.com.

Vacationing with the Stars: Star Parties, Eclipse Trips, and Telescope Motels

An astronomy vacation is a treat for the mind and a feast for the eyes. Plus, traveling with the stars is often cheaper than a conventional holiday. You don't have to visit the hottest tourist destinations to keep up with your snooty neighbors. You can have the experience of a lifetime and come back raving about what you saw and did, not just what you ate and spent.

You can, however, blow big bucks on one type of astronomy vacation: the eclipse cruise. But if you like ocean cruises, taking one to an eclipse doesn't cost any more than similar voyages that have no celestial rewards. And bargain-basement eclipse tours are available, too. Star parties and telescope motels are additional options; I cover all the bases in the following sections. Pack your bags and have the neighbors watch your dog!

Party on! Attending star parties

Star parties are outdoor conventions for amateur astronomers. Hundreds of them set up their telescopes (some homemade and some not) in a field and people take turns skywatching. (Be prepared to hear plenty of "Oohs" and "Ahs.") The best homemade telescopes and equipment receive prizes from the party organizers. If it rains in the evening, partygoers can watch slide shows in a nearby hall or a big tent. Arrangements vary, but often some attendees camp in the field and others rent inexpensive cabins or commute from nearby motels. Star parties usually last several days and nights (sometimes as much as a week). They attract a few hundred to a few thousand (yes, thousand!) telescope makers and amateur astronomers. And the larger star parties have Web sites with photos of previous events and details on coming attractions.

The leading star parties in the United States include

- Stellafane, in Vermont (www.stellafane.com)

- Texas Star Party, where you can be alone with the stars in the Lone Star State (www.texasstarparty.org)

- RTMC Astronomy Expo, where you can get high on the stars at an elevation of 7,600 feet in California's San Bernadino mountains (www.rtmc astronomyexpo.org)

- Enchanted Skies Star Party, out in the desert in New Mexico (www.socorro-nm.com/starparty)

- Nebraska Star Party, with a very dark site in the Sandhills (www.nebraska starparty.org)

In the long run, I recommend that you visit at least one of the above leading star parties, but in the meantime, you can find a star party or other astronomy event near you by using Sky & Telescope's online Event Calendar at skyandtelescope.com/resources/calendar.

To the path of totality: Taking eclipse cruises and tours

Eclipse cruises and tours are planned voyages to the places where you can view total eclipses of the sun. Astronomers can calculate long in advance when and where an eclipse will be visible. The locations where you can see the total eclipse are limited to a narrow strip across land and sea, the *path of totality.* You can stay home and wait for a total eclipse to come to you, but you may not live long enough to see more than one, if any. So if you're the impatient stargazing type, you may want to travel to the path of totality, because the next total eclipse of the sun visible from the continental United States isn't until 2017.

Recognizing reasons to book a tour

If an eclipse is visible within easy driving distance, you don't need to sign up for a tour. And if you're an experienced domestic and international traveler, you can arrange to go on your own to the path of totality for a distant eclipse. But consider this fact: Expert meteorologists and astronomers identify the best viewing locations years in advance. More often than not, these places aren't vast metropolises that offer huge numbers of vacant accommodations. You have to travel to random spots on the globe. After experts tab a spot as a prime location for a coming eclipse, tour promoters and savvy individuals book all or most of the local hotels and other facilities years in advance. Johnnies-or-Janies-come-lately, especially those who travel on their own, may be out of luck.

A tour promoter usually engages a meteorologist and a few professional astronomers (sometimes even me!). So you have the benefit of a weatherperson to make last-minute decisions on moving the group's observing site to a place with a better next-day forecast, an astronomer to show you the safest methods to photograph the eclipse, and usually another lecturer who tells old eclipse tales and reports on the latest discoveries about the sun and space. And, if you travel out into the boondocks, at least one of the experts carries a Global Positioning System (GPS) receiver to verify that the group is at the desired location.

On the night following the eclipse, everyone shows their videotapes of the darkening sky, the birds coming to roost, a klutz knocking over his telescope at the worst possible time, and the excited crowd saying "Wow" and "Hurray." And, of course, they replay the eclipse — over and over.

If these details haven't convinced you to take an eclipse tour for your viewing pleasure, consider this: A group tour to a foreign destination is almost always cheaper than going on your own (and more satisfying than waiting 15 years).

You can do it! Participating in scientific research

You can make your astronomy hobby beneficial and fun by joining national and worldwide efforts to gather precious scientific data. You may have only a pair of binoculars compared to Keck Observatory's two 10-meter-wide (400-inch-wide) telescopes, but if Mauna Kea has a cloudy day, Keck can't see anything. And if a spectacular fireball shoots over your hometown, you may be the only astronomer to see it.

Secret U.S. Department of Defense satellites and an amateur moviemaker vacationing at Glacier National Park recorded one of the most spectacular and interesting meteors of all time. A clip from that home movie appears in just about every scientific documentary about meteors, asteroids, and comets that appears on television. It pays to be in the right place at the right time. And some day, you may be in that position.

Join other amateur astronomers and enjoy the projects I recommend throughout this book. You can do these activities on your own, but you may find it easier to compare notes with someone who has experience, so ask around at your local astronomy club.

Examining the advantages of cruising

An eclipse cruise is better than a tour but more expensive. At sea, the captain and navigator have "two degrees of freedom." When the meteorologist says "head southwest down the path of totality for 200 miles" on the night before the eclipse (because of a best-case forecast for a cloud-free location at eclipse time), the ship can follow those instructions. But on land, you have to keep the bus on the road, and a road may not go in the direction you need it to go. If the bus does have a suitable road, you run into thousands of eclipse-trippers trying to get a better site at the last minute because of the weather change. On the cruise, you can leave the steering to the crew, recline in your deck chair, sip a piña colada, have your camera ready, and wait for totality.

I've viewed many eclipses, and my experience is that if you stay on the ground, you get to see totality about half the time. But if you travel on an ocean liner, you never miss.

Making the right decision

You can find advertisements for eclipse tours and cruises in astronomy magazines and in many science and nature magazines, and you can also book through travel agents. Clubs, fraternal organizations, and college alumni societies often sponsor the cruises you find.

Here are some ways you can choose the right tour or cruise for you:

✔ **Consult current and back issues of astronomy magazines.** Most run articles about the viewing prospects for a solar eclipse a few years in advance. Get their expert recommendations about the best viewing sites.

✔ **Check out the advertisements from travel operators.** Which tours and cruises go to the best places? Get brochures from travel agents, tour promoters, and cruise lines. Often, promoters list previous successful eclipse trips, indicating a level of experience.

Motoring to telescope motels

Telescope motels are resorts where the attractions are the dark sky and the opportunity to set up your own telescope in an excellent viewing location. They usually have telescopes that you can use, perhaps at an additional fee.

Some telescope motels worth visiting in the United States are

✔ Star Hill Inn, at an altitude of 7,200 feet in Sapello, New Mexico, is a pioneer in resort lodgings for astronomers. (www.starhillinn.com)

✔ Skywatcher's Inn, in Benson, Arizona, features a 20-inch telescope — you can stay in the Galaxy Room, with its own planetarium and dome ceiling. (www.skywatchersinn.com)

✔ The Observer's Inn is located in the historic gold-mining town of Julian, California. Although the inn suffered damage from a forest fire, the guest quarters should reopen in the first quarter of 2005. (www.observers inn.com)

✔ StarGazers Inn & Observatory, in Big Bear Lake, California, is a luxurious bed and breakfast, with telescopes and binoculars to use and candlelit hot tubs to relax in "after sky." (www.stargazersinn.com)

Overseas, two telescope motels worth visiting are

✔ COAA (Centro de Observação Astronómica no Algarve), in southern Portugal, with at least four telescopes and three guest suites (www.coaa.co.uk/index.html)

✔ Fieldview Guest House, a bed and breakfast with six telescopes, off the beaten track in East Anglia, United Kingdom (www.fieldview.net)

Chapter 3

The Way You Watch Tonight: Terrific Tools for Observing the Skies

In This Chapter

▶ Familiarizing yourself with the night sky

▶ Observing objects with the naked eye

▶ Putting binoculars and telescopes to good use

▶ Setting up a surefire observation plan

*I*f you've ever gone outside and looked at the night sky, you've been stargazing — observing the stars and other objects in the sky. Naked-eye observation can distinguish colors and the relationships between objects — like finding the North Star by using "pointer stars" in the Big Dipper.

From naked-eye observation, you can take a short step up by adding optics to see fainter stars and view objects with greater detail. First try binoculars, and then graduate to a telescope. Next thing you know, you're an astronomer!

But I'm getting ahead of myself. First, you need to take some quiet looks at the cosmos and see the beauty and mystery for yourself. You can use three basic tools — at least one of which you already own.

Whether you use your eyes, a pair of binoculars, or a telescope, each method of observation is best for some purposes:

✔ The human eye is ideal for watching meteors, the aurora borealis, or a conjunction of the planets (when two or more planets are close to each other in the sky) or of a planet and the moon.

✔ Binoculars are best for observing bright variable stars, which are too far from their *comparison stars* (stars of known constant brightness used as a reference to estimate the brightness of a star that varies in brightness) to see them together through a telescope. And binoculars are wonderful for sweeping through the Milky Way and viewing the bright nebulae and star clusters that dot it here and there. Some of the brighter galaxies — such as M31 in Andromeda, the Magellanic Clouds, and M33 in Triangulum — look best through binoculars.

✔ You need a telescope to get a decent look at most galaxies and to distinguish the members of close double stars, among many other uses. (A double star consists of two stars that appear very close together; they may or may not be near each other in space, but when they truly are together, they form a binary star system.)

In this chapter, I cover these observational tools, provide you with a quick primer on the geography of the night sky, and give you a handy plan for delving into astronomy. Before long, you'll be observing the skies with ease.

Seeing Stars: A Sky Geography Primer

When viewed from the Northern Hemisphere, the whole sky seems to revolve around the North Celestial Pole. Close to the NCP is the North Star (also called Polaris), a good reference point for stargazers because it always appears in almost exactly the same place in the sky, all night long (and all day long, but you can't see it then).

In the following sections, I show you how to familiarize yourself with the North Star and give you a few facts about constellations.

As the earth turns . . .

Our Earth turns. The Greek philosopher Heraclides Ponticus proclaimed that concept in the fourth century B.C. But people doubted Heraclides's observations, because folks thought that if his theories were true, they should feel dizzy like riders of a fast merry-go-round or a whirling chariot. They couldn't imagine a turning Earth if they couldn't physically feel the effects. Instead, the ancients thought the sun raced around Earth, making a complete revolution every day. (They didn't feel the effects of Earth's rotation, nor do you and I, because they're too small to notice.)

Proof of Earth's turning, or *rotation,* didn't come until 1815, more than two millennia after Heraclides (researchers didn't have much government funding back then, so progress was slow). The proof came from a big French swinger: a heavy metal ball suspended from the ceiling above the floor of the Pantheon (a church) in Paris on a 200-foot wire. The ball is called a *Foucault pendulum,* after the French physicist who came up with the plan. If you kept an eye on the pendulum as it swung back and forth all day, you could see that the direction taken by the swinging ball across the floor gradually changed, as though the floor was turning underneath it. And it was; the floor turned with the earth.

If you're not convinced that the earth turns, or if you just like to watch swingers, you can see a Foucault pendulum in the Science Museum of Virginia in Richmond (www.smv.org/info/foucaultEX.htm). If you believe, however, you can just sip your favorite beverage as you enjoy a sunset.

As I explain in Chapter 1, the rotation of Earth around its axis makes the stars and other sky objects appear to move across the sky from east to west. In addition, the sun moves across the sky during the year on a circle called the *ecliptic.* (If you could see the stars in the daytime, you'd note the sun moving to the west across the constellations, day by day.) The ecliptic is inclined by 23.5 degrees to the celestial equator, the same angle by which the axis of Earth is tilted from the perpendicular to its orbital plane.

The planets stay close to the ecliptic as they move throughout the year. They move systematically through 12 constellations located on the ecliptic, which collectively are called the *Zodiac:* Aries, Taurus, Gemini, Cancer, Leo, Virgo, Libra, Scorpius, Sagittarius, Capricorn, Aquarius, and Pisces. (Actually, a 13th constellation intersects the ecliptic — Ophiuchus — but in ancient times, it wasn't included with the original 12.)

Earth's steady progression along its orbit of the sun results in a different appearance of the night sky over the course of the year. The stars aren't located in the same places with respect to the horizon throughout the night or throughout a year. The constellations that appeared high in the sky at dusk a month ago are lower in the west at dusk now. And if you view the constellations that loom low in the east just before dawn, you preview what you'll see at midnight in a few months.

To keep track of the constellations, use the sky maps that come monthly in astronomy magazines such as *Sky & Telescope* or *Astronomy.* (For more about magazines, check out Chapter 2.) Your daily newspaper may also contain maps of the evening skies, or you can get an inexpensive *planisphere,* which features a rotating wheel in a frame — with a hole cut into it that represents the limits of your view — that signifies the night sky. You can adjust it according to the time and date. Hardin Optical, in Bandon, Oregon, sells a variety of planispheres, including its own "Hardin Optical Star and Planet Guide" and also "David H. Levy's Guide to the Stars," which is available in both English- and Spanish-language editions. Hardin's Web site is www.hardinoptical.com. When you log on to the site, enter "planisphere" in the search window.

. . . *keep an eye on the North Star*

Anyone can walk outside on a clear night and see some stars. But how do you know what you're seeing? How can you find it again? What should you watch for?

One of the most time-honored ways of getting familiar with the night sky if you live in the Northern Hemisphere is to become familiar with the North

Star, or Polaris, which barely moves. After you identify which way is north, you can orient yourself to the rest of the northern sky, or to anywhere else for that matter. In the southern sky, you need to find the bright stars Alpha and Beta Centauri (a Southern Hemisphere planisphere or a simple star map can help), which point the way to the Southern Cross.

You can easily find the North Star by using the Big Dipper in the constellation Ursa Major (see Figure 3-1). The Big Dipper is one of the most easily recognized sky patterns. If you live in the continental United States, you can see it every night of the year.

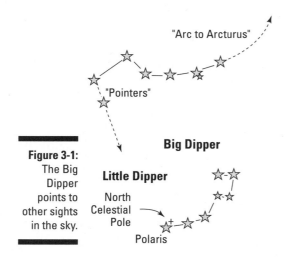

Figure 3-1:
The Big
Dipper
points to
other sights
in the sky.

The two brightest stars in the Big Dipper, Dubhe and Merak, form one end of its bowl and point directly to the North Star. The Big Dipper helps you locate the bright star Arcturus, in Bootes, too: Just imagine a smooth continuation of the curve of the Dipper's handle, as shown in Figure 3-1.

The stars close to Polaris never set below the horizon at most latitudes in North America, which makes them *circumpolar stars:* They appear to circle around Polaris. Ursa Major is a circumpolar constellation seen from almost all the Northern Hemisphere. The circumpolar area of the sky depends on your latitude. The closer you live to the North Pole, the more of the sky is circumpolar. And in the Southern Hemisphere, the farther south your location, the greater the circumpolar part of the sky.

Although not circumpolar for most viewers, Orion is a very distinctive constellation visible in the Northern Hemisphere's winter evening sky, with the three stars that make up its belt pointing to Sirius in Canis Major and Aldebaran in Taurus. Orion also contains the first-magnitude stars Betelgeuse and Rigel, two brilliant beacons in the sky (see Figure 3-2). For more on the magnitude of stars, head to Chapter 1.

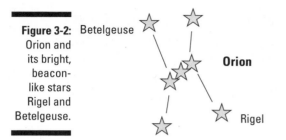

Figure 3-2:
Orion and
its bright,
beacon-
like stars
Rigel and
Betelgeuse.

You can become friends with the night sky by looking at this book's constellation maps (check out Appendix B) and checking it out with your own eyes. Just like becoming familiar with the streets of your city helps you find your way faster, knowing the constellations helps you set your sights on the sky objects you want to observe. Gaining sky knowledge also assists you in tracking the appearance of the stars and their movements as you maneuver through a nightly session.

Beginning with Naked-Eye Observation

If you don't already know the compass directions in your area, take the time to familiarize yourself with them. You need to know north from south and east from west. After you get your bearings, you can use the weekly sky highlights from the *Sky & Telescope* Web site (skyandtelescope.com) or the picture that fills your screen when you run a desktop planetarium program to orient yourself to the brightest stars and planets. (For more about astronomy resources, see Chapter 2.) When you recognize the bright stars, you have an easier time picking out the patterns of slightly fainter ones all around them.

Table 3-1 lists some of the brightest stars you can see in the night sky, the constellations that contain them, and their magnitudes (the measure of their brightness — see Chapter 1). Many of them are visible from the continental United States and Canada. Some you can only see from southern latitudes, so bright stars you can't see in the United States may be prominent sights for Australians. See Chapter 11 for information on spectral class, which is an indication of the color and temperature of a star. (Spectral class B stars, for example, are white and rather hot, and M stars are red and relatively cool.)

Start your observations by consulting a star map or desktop planetarium and seeing how many of the brightest stars you can locate at night. After you do, try identifying some of the dimmer stars in the same constellations. And, of course, keep your eye out for the bright planets: Mercury, Venus, Mars, Jupiter, and Saturn (which I cover in Chapters 6 and 8).

Table 3-1	The Brightest Stars as Seen from Earth		
Common Name	Apparent Magnitude	Constellation Designation	Spectral Class
Sirius	−1.5	α Canis Majoris	A
Canopus	−0.7	α Carinae	A
Rigil Kentaurus	−0.3	α Centauri	G
Arcturus	−0.04	α Bootes	K
Vega	0.03	α Lyrae	A
Capella	0.1	α Aurigae	G
Rigel	0.1	β Orionis	B
Procyon	0.4	α Canis Minoris	F
Achernar	0.5	α Eridani	B
Betelgeuse	0.5	α Orionis	M
Hadar	0.6	β Centauri	B
Acrux	0.8	α Crucis	B
Altair	0.8	α Aquilae	A
Aldebaran	0.9	α Tauri	K
Antares	1.0	α Scorpii	M
Spica	1.0	α Virginis	B
Pollux	1.1	β Geminorum	K
Fomalhaut	1.2	α Piscis Austrini	A
Deneb	1.3	α Cygni	A

In winter and summer, the Milky Way runs high in the sky at most locations in the United States. If you can recognize the Milky Way as a wide, faintly luminous band across the sky, you have at least a pretty fair observing site.

The most important step in observing with the naked eye is to shield your vision from interfering lights. If you can't get to a dark place in the country, find a dark spot in your backyard or possibly on the roof of a building. You won't eliminate the light pollution high up in the sky that results from the collective lights of your city, but trees or the wall of a house can prevent nearby lights, including streetlamps, from shining in your eyes. After 10 or 20 minutes, you can see fainter stars; you're getting "dark adapted."

Watching the bright comet Hyakutake in 1996 from a small city in the Finger Lakes of northern New York, I found that walking around the corner of a building to put myself in its shadow made an enormous difference in the visibility of the comet.

Ideally, you want a site with a good horizon and only trees and low buildings in the distance, but finding that kind of location is next to impossible in a major urban area.

If you fail to find a site with a good horizon in all directions, the most important horizon is the southern one. You make most observations in the Northern Hemisphere while facing roughly south (with the east to the left and west to the right). As you face south, the stars rise to your left and set to your right. If you live in or visit the Southern Hemisphere, reverse this procedure and face north.

Always have a watch, a notebook, and a dim or red flashlight to use for recording what you see. Some flashlights come with a red bulb, or you can buy red cellophane from a greeting card store to wrap around the lamp. After you become dark adapted, white light reduces your ability to see the faint stars, but dim red light doesn't hurt your dark adaptation.

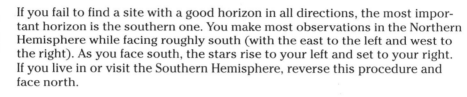

How bright is bright?

I talk about magnitude in Chapter 1, but it helps to know that astronomers can define magnitude in different ways for different purposes:

✔ *Absolute magnitude* is the brightness of a sky object as seen from a standard distance of 32.6 light-years. Astronomers consider it the "true" magnitude of the object.

✔ *Apparent magnitude* is how bright an object appears from Earth, which is usually different from its absolute magnitude, depending on how far away from Earth the sky object is located. A star closer to Earth may appear brighter than one farther away, even if its absolute magnitude is fainter.

✔ *Limiting magnitude* is related to how clear and dark the sky is at the time of observation — a very bright object may be invisible if many clouds hang overhead, for example. Limiting magnitude is especially important in meteor and deep sky observations. On a clear dark night, the limiting magnitude may be 6 at the zenith, but in the city the limiting magnitude may be only 4.

Star charts depict the apparent magnitudes of the stars to simulate their appearances in the sky.

Using Binoculars or a Telescope for a Better View

As with any new hobby, you should gain some experience and research what's available before you start buying expensive equipment. You shouldn't buy a telescope until you've seen several telescopes of different types in action and discussed them with other observers. In the following sections, I offer my advice for choosing the right binoculars or telescope for you.

Binoculars: Sweeping the night sky

Owning a good pair of binoculars is a must. Buy or borrow a pair before you get a telescope. Binoculars are excellent for many kinds of observation, and if (sigh) you give up on astronomy, you can still use them for many other purposes. Just don't leave your neighbor's pair in your garage.

Binoculars are great for observing variable stars, searching for bright comets and novae, and sweeping the sky just to enjoy the view. You may never discover a comet yourself, but you'll certainly want to view some of the brighter ones as they appear. Nothing works better for this purpose than a good pair of binoculars.

The following sections cover the way binoculars are specified according to their capabilities, and I show you the steps to take as you figure out what kind of binoculars to buy. Figure 3-3 takes you inside a pair of binoculars.

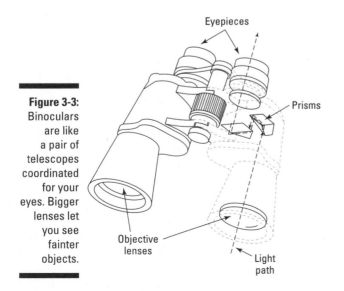

Figure 3-3: Binoculars are like a pair of telescopes coordinated for your eyes. Bigger lenses let you see fainter objects.

Deciphering the numbers on binoculars

Binoculars come in many sizes and types, but each pair of binoculars is described by a numerical rating — 7 x 35, 7 x 50, 16 x 50, 11 x 80, and so on. (Note that the ratings read "7 by 35, 7 by 50, 16 by 50, and 11 by 80." Don't say "7 times 35.") Here's how to decode these ratings:

- The first number is the optical magnification. A 7 x 35 or 7 x 50 pair of binoculars makes objects look seven times larger than they do to the naked eye.

- The second number is the *aperture,* or diameter, of the light-collecting lenses (the big lenses) in the binoculars, measured in millimeters. An inch is about 25.4 millimeters; thus, 7 x 35 and 7 x 50 binoculars have the same magnifying power, but the 7 x 50 pair has bigger lenses that collect more light and show you fainter stars than the 7 x 35 pair.

You should also keep the following considerations in mind:

- Bigger binoculars reveal fainter objects than smaller ones do, but they weigh more and are harder to hold and point steadily toward the sky.

- Higher magnification binoculars, such as 10 x 50 and 16 x 50, show objects with greater clarity, provided that you can hold them steady enough, but they have smaller fields of view, so finding celestial targets is harder than with lower magnification binoculars.

- Giant binoculars — 11 x 80, 20 x 80, and on up — are heavy and hard to hold steady; many people can't use them without a tripod or stand. The very biggest, 40 x 150, must be used with a stand.

- Many intermediate sizes are available, such as 8 x 40 or 9 x 56.

Here's my opinion: 7 x 50 is the best size for most astronomical purposes and certainly the best size to start with. If you purchase binoculars much smaller than 7 x 50, you really equip yourself for birdwatching rather than astronomy. Buy a much larger size than 7 x 50 and you may be investing in a white elephant that you'll rarely use.

Making sure your binoculars are right for you

First and foremost, you shouldn't buy binoculars unless you can return them after a trial run. Here's how to make the basic check to determine whether a pair of binoculars is worth keeping:

- The image should be sharp across the field of view when you look at a field of stars.

- You should have no difficulty focusing the binoculars for your eyesight, with a separate adjustment for at least one of the eyepieces (the small lenses next to your eyes when you look through the binoculars).

> ✔ When you adjust the focus, it should change smoothly. Stars' images should be sharp points when in focus and circular in shape when not.
>
> ✔ Special transparent coatings are deposited on the objective lenses (large lenses) of many binoculars. This feature, called *multi-coating,* results in a clearer, more contrasting view of star fields.

Good binoculars are sold in optical and scientific specialty stores. Some large camera stores have decent binocular selections. But I suggest that you avoid the department stores. You may get low-grade merchandise in some department stores or pay exorbitant prices for fancy binoculars in others. And you can bet that the salespeople peddling them know less than you do.

You can pay hundreds of dollars or even a few thousand bucks for a good pair of 7 x 50 binoculars, but if you shop around, you can find a perfectly adequate pair for $120 or less. (A military surplus store is an excellent place to look.) Used binoculars are often a good deal.

Many astronomers buy their binoculars from specialty retailers and manufacturers that advertise in astronomy magazines and on the Web (see Chapter 2 for more about these resources). If you must order by mail (or from the Web), ask experienced amateurs whom you meet at an astronomy club or a staff member at a planetarium to recommend a dealer.

Reputable makers of binoculars include Bausch & Lomb, Bushnell, Canon, Celestron, Fujinon, Leica, Meade, Nikon, Orion, and Pentax. Some high-end Canon binoculars have image stabilization, a high-tech feature that makes the image much steadier. They come in handy on a boat rocking at sea.

Telescopes: When closeness counts

If you want to look at the craters on the moon or the surfaces and cloud decks of the planets, you need a telescope. The same advice goes for observing faint variable stars or viewing galaxies and the beautiful small glowing clouds called *planetary nebulae,* which have nothing to do with planets (see Chapters 11 and 12).

Before you view the sun or any object crossing in front of the sun, however, be sure to read the special instructions in Chapter 10 to protect your eyes and avoid going blind!

The following sections cover telescope classifications, mounts, and shopping tips to find the best telescope for your needs.

Focusing on telescope classifications

Telescopes come in three main classifications:

✔ Refractors use *lenses* to collect and focus light (see Figure 3-4). In most cases, you look straight through a refractor.

✔ Reflectors use *mirrors* to collect and focus light (see Figure 3-5). Reflectors come in different types:

  • In a *Newtonian* reflector, you look through an eyepiece at right angles to the telescope tube.

  • In a *Cassegrain* telescope, you look through an eyepiece at the bottom.

  • A *Dobsonian* reflector gives you the most *aperture* (or light gathering power) for your money, but you may have to stand on a stool or a ladder to look through it. Dobsonians tend to be larger than other amateur telescopes (because large Dobsonians are more affordable), and the eyepiece is up near the top.

✔ *Schmidt-Cassegrains* and *Maksutov-Cassegrains* use both mirrors and lenses. These models are more expensive than reflectors with comparable apertures.

Many varieties are available within these general telescope types. And every telescope used for amateur purposes is equipped with an *eyepiece,* which is a special lens (actually a combination of lenses mounted together as a unit) that magnifies the focused image for viewing. When you take photographs, you generally don't use an eyepiece.

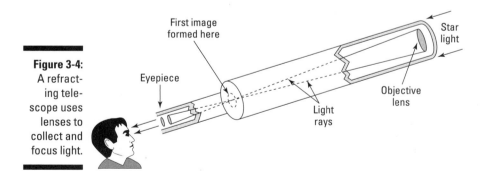

Figure 3-4: A refracting telescope uses lenses to collect and focus light.

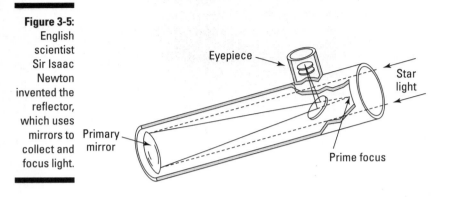

Figure 3-5:
English scientist Sir Isaac Newton invented the reflector, which uses mirrors to collect and focus light.

Just like a microscope or a camera with interchangeable lenses, you can use interchangeable eyepieces with almost any telescope. Some companies don't make telescopes at all, specializing instead in making eyepieces that conform to many different telescopes.

Beginners usually buy the highest magnification eyepieces they can, which is a great way to waste your money. I recommend low- and medium-power eyepieces, because the higher the power of the piece, the smaller the field of view, making it tougher to track faint (and possibly even bright) targets. For a small telescope, observation is usually best with eyepieces rated around 25x or 50x, not 200x or more. (The *x* stands for "times," as in 25 times bigger than what you see with the naked eye.) If you see a telescope advertised for its "high power," the advertisers may be trying to sell mediocre goods to unwitting buyers. And if a salesperson touts the high power of a telescope, patronize another store.

What limits your view of fine details with a small telescope isn't the power of the eyepiece. The turbulence in the atmosphere or even the shaking of the telescope in the breeze affects your view.

Examining telescope mounts

Telescopes are generally mounted on a stand, a tripod, or a pier in one of two ways:

- ✔ With an *alt-azimuth mount,* you can swivel the telescope up and down and side to side — in altitude (the vertical plane) and azimuth (the horizontal plane). You need to adjust the telescope on both axes to compensate for the motion of the sky as the earth rotates. Dobsonian reflectors always use alt-azimuth mounts.

- ✔ With the more expensive *equatorial mount,* you align one axis of the telescope to point directly at the Celestial North Pole, or for Southern Hemisphere viewers, the Celestial South Pole. After you spot an object, simply turning the telescope around the polar axis keeps it in view. Be sure to polar align the scope for each viewing session.

Coloring your universe

What do you see when you look at a sky object through your binoculars or telescope? Do you see glorious stars, planets, and sky objects in brilliant colors, as shown in the photos in the color section of this book? Not!

Sorry for the tease, but you're likely to see most stars and sky objects in pale colors. Most stars look white or off-white by eye, binoculars, or telescope — yellowish rather than yellow, for example. The colors are most vivid when adjacent stars are in strong contrast, as found in telescopic views of some double stars.

Most photos of sky objects have been color enhanced, traditionally described as having *false color.* Astronomers don't use false color to gussy up the universe, which is bright and beautiful all on its own, thank you very much. Nor is

it meant to give a false impression of the deep sky. In fact, the enhancement furthers the search for truth, much like a stain on medical slides brings out the detail in the cells and helps identify physical differences and relationships.

Depending on the method of observation and presentation, photos of the same object can be strikingly different. But they all tell scientists about differences in the structure of the object, what substances it may contain, and what dynamic processes are taking place. Also, many astronomical images are obtained in forms of light invisible to the human eye (like ultraviolet, infrared, and X-ray), so astronomers use false color in this absence of any recognizable color.

An alt-azimuth mount is usually steadier and easier for a beginner to use, but an equatorial mount is better for tracking the stars as they rise and set.

The objects you see in the telescope are usually upside down, which isn't the case with binoculars. Of course, it doesn't really make much difference in the viewing, but just know that top and bottom are reversed when you view through a telescope. Adding a lens to rotate the image reduces the light coming to the scope and thus dims the image slightly. When viewed through an equatorially mounted telescope, a star field maintains the same orientation as it rises and sets. But with an alt-azimuth mount telescope, the field rotates during the night, so the stars on top wind up on the side.

Shopping for telescopes the smart (and economical) way

A cheap, mass-manufactured telescope, often called a *drugstore* or *department store* telescope, is usually a waste of money. And it still costs more than a hundred or maybe several hundred dollars.

A good telescope, bought new, may run you the better part of $1,000, and you certainly can pay more. But you can find alternatives:

Staying safe when you view the sun

Taking even the briefest peek at the sun through a telescope, binoculars, or any other optical instrument is very dangerous unless the device is equipped with a solar filter made by a reputable manufacturer specifically for viewing the sun. And the filter must be properly mounted on the telescope, not jury-rigged.

You must also use a solar filter when you view planets that cross the disk of the sun. (Such planets are said to be in *transit*.) Viewing an object against the sun requires the use of protective viewing techniques, because you're also looking at the sun. If you have a Newtonian reflector, a Dobsonian reflector, or a refractor, you can try using projection. See Chapter 10 for specifics about solar viewing and protecting your eyes.

✔ Used telescopes are often for sale, via ads in astronomy magazines or in the newsletters of local astronomy clubs. If you can inspect and test a used telescope and you find what you like, buy it! A well-maintained telescope can last for decades.

✔ In many areas, amateurs can observe with the larger telescopes operated by astronomy clubs, planetariums, or public observatories.

The technology of amateur telescopes is advancing at a rapid pace, and a former astronomer's dream can be today's obsolete equipment. Quality and capabilities are going up and price is going down.

Generally speaking, a good refractor gives better views than a good reflector that has the same aperture or telescope size. Aperture or telescope size refers to the diameter of the main lens, mirror, or, in a more complicated telescope, the size of the unobstructed portion of the optics. But a good refractor is much more expensive than a reflector.

The Maksutov-Cassegrains and Schmidt-Cassegrains are good compromises between the low cost of a reflector and the high performance of a refractor. For many astronomers, these models are the preferred telescopes.

As of early 2005, one of the best small telescopes was the Meade ETX-90PE, a mightily upgraded version of the older ETX-90. Its aperture is 3.5 inches, almost the smallest size of any telescope that you should start with. (If you find a good instrument at a good price from a 2.5-inch aperture on up, especially in a refractor, consider it for purchase.)

The basic Meade ETX-90PE lists at $695 and comes with an Autostar computerized controller and a tripod (each of which used to cost extra). This instrument automatically points at almost any object you specify, if that object is in view from your location at that time. The Autostar can even find moving objects, such as planets, based on stored information, and it's equipped to give you a "tour" of the best sights in the sky, selected with no input from you.

You definitely don't want to spend this much money until you see the telescope in action at an astronomy club observing meeting or a star party (see Chapter 2). But the price is no more than you pay for a fine camera and an accessory lens or two. You can find larger telescopes for less money — check the ads in current issues of astronomy magazines — but you have to invest much more effort in learning to use them effectively.

Some brand-name telescopes are sold only through authorized dealers who tend to have expert knowledge. But take their advice with just a wee bit of salt, especially when they carry several competing lines of telescopes and make their own too.

Key Web sites to browse for telescope product information are

✔ Celestron, for many years the favorite manufacturer for thousands of astronomers (www.celestron.com)

✔ Meade Instruments Corporation (www.meade.com)

✔ Orion Telescopes & Binoculars (www.telescope.com)

On each of these Web sites, you can find the instruction manuals for many of the telescopes that they sell. Consider taking a look at the manual before you buy a telescope so you know whether it will be helpful when you run into problems.

Good seeing gone bad

Turbulence in the atmosphere affects how well you can see the stars. It makes the stars seem to twinkle. The term *seeing* describes the conditions of the atmosphere relating to steadiness of the image — *good seeing* is when the air is stable and the image holds steady. You often have better seeing late at night when the heat of the day has dissipated. When seeing is bad, the image tends to "break up," and double stars blur together in telescopic views. The stars always twinkle most close to the horizon where the seeing is worst.

The warmth of a telescope brought out from a heated home into the cool night air causes some bad seeing. You should wait a while for the telescope to cool down; your viewing will improve. Situations vary, but 30 minutes is usually enough to make a significant difference in the viewing.

Planning Out Your Dip into Astronomy

I recommend that you get into the astronomy hobby gradually, investing as little money as possible until you're sure about what you want to do. Here's a plan for acquiring both basic skills and the needed equipment:

1. **If you have a late-model computer, invest in an inexpensive planetarium program. Start making naked-eye observations at dusk on every clear night and before dawn if you're an early riser.**

   To plan your observations of planets and constellations, rely on the weekly sky scenes at the *Sky & Telescope* Web site (skyandtelescope.com). If you don't have a suitable computer, plan your observations based on the monthly sky highlights in *Astronomy* or *Sky & Telescope* magazine.

2. **After a month or two of familiarizing yourself with the sky and discovering how much you enjoy it, invest in a serviceable pair of 7 x 50 binoculars.**

3. **As you continue to observe the bright stars and constellations, invest in a star atlas that shows many of the dimmer stars, as well as star clusters and nebulae.**

   Norton's Star Atlas and Reference Handbook (edited by Ian Ridpath, 20th edition, by PI Press, 2004) has been a leading product for generations. But see what else is available in magazine ads and at planetarium and science museum gift shops. Compare scenes in your star atlas with the constellations that you're observing; the atlas shows their RAs and Decs (see Chapter 1 for info about RAs and Decs). Eventually, you'll start to develop a good feel for the coordinate system.

4. **Join an astronomy club in your area, if at all possible, and get to know the folks who have experience with telescopes. (See Chapter 2 for more about finding clubs.)**

5. **If all goes well, and you want to continue in astronomy — as I bet you will — invest in a well-made, high-quality telescope in the 2.5- to 4-inch size range.**

   Study the telescope manufacturer Web sites earlier in this chapter or send for catalogs advertised in astronomy magazines. Better yet, talk to experienced astronomy club members if you can.

If you find that you enjoy astronomy as much as I think you will, after a few years you should consider moving up to a 6-inch or 8-inch telescope. It may be harder to use, but you'll be ready to master it after you have some experience. Equipped with a larger telescope, you can see many more stars and other objects.

Chapter 4

Just Passing Through: Meteors, Comets, and Artificial Satellites

In This Chapter

▶ Finding fast facts about meteors, meteoroids, and meteorites

▶ Following the heads or tails of comets

▶ Spotting artificial satellites

See a moving object in the daytime sky? You probably know whether it's a bird, a plane, or Superman. But in the night sky, can you distinguish the light of a meteoroid from that of an Iridium satellite? And among objects that move slowly but perceptibly across the starry background, can you tell a comet from an asteroid?

This chapter defines and explains many of the objects that sweep across the night sky. (The sun, moon, and planets move across the sky, too, but in a more stately procession. I focus on them in Parts II and III.) When you know these night visitors, you can look forward to enjoying them all.

Meteors: Wishing on a Shooting Star

No astronomy term is misused more often than the word "meteor." Amateur astronomers, and even scientists, are quick to spurt out "meteor" when meteoroid or meteorite is the accurate term. Here are the correct meanings:

- A *meteor* is the flash of light produced when a naturally occurring small, solid object (a meteoroid) enters Earth's atmosphere from space; people often call meteors "shooting stars" or "falling stars."

- A *meteoroid* is a small, solid object in space, usually a fragment from an asteroid or comet, orbiting the sun. Some rare meteoroids are actually rocks blasted off Mars and Earth's moon.

- A *meteorite* is a solid object from space that has fallen to the surface of Earth.

If a meteoroid runs into Earth's atmosphere, it may produce a meteor bright enough for you to see. If the meteoroid is big enough to hit the ground instead of disintegrating in midair, it becomes a meteorite. Many people hunt for and collect meteorites because of their value to scientists and collectors.

The two main kinds of meteoroids have different places of origin:

- *Cometary meteoroids* are fluffy little dust particles shed by comets.
- *Asteroidal meteoroids,* which range in size from microscopic particles to boulders, are literally chips from asteroids — the so-called minor planets — which are rocky bodies that orbit the sun (and which I describe in Chapter 7).

When you go to a science museum and see a meteorite on display, you're examining an asteroidal meteoroid that fell to the earth (or, in rare cases, a rock that fell after being knocked off the moon or Mars by a larger impacting body). It may be made of stone, iron (actually an almost rustproof mixture of nickel and iron), or both. Showing rare simplicity (for once), scientists call these meteorite types, respectively, *stony, iron,* and *stony-iron meteorites.*

In the following sections, I cover three types of meteors: sporadic meteors, fireballs, and bolides. I also give you the scoop on meteor showers.

Comb yourself for space dust

If an astronomer finds a *micrometeorite* (a meteorite so small that you must view it through a microscope), it may be a particle that began as a cometary meteoroid, or it may be a very small asteroidal meteoroid.

Micrometeorites are so small that they don't create enough friction to burn up or disintegrate in the atmosphere, so they sift slowly down to the ground. Chances are, you have one or two pieces of this space dust in your hair right now, but the dust is almost impossible to identify because it would be lost among the millions of other microscopic particles on your head (no offense).

Scientists obtain micrometeorites by flying ultra-clean collecting plates on high-altitude jet aircraft. And they drag magnetized rakes, which pick up micrometeorites made of iron, through the mud on the sea floor.

In January 2004, NASA's Stardust space probe flew past Comet Wild-2 (a small comet that passes inside the orbit of Mars once every six years or so and is thus fairly easy to reach with a probe) and collected some of the comet's dust. Finishing its "Wild ride," Stardust is due to drop the dust sample off on Earth by parachute in 2006. You can see where Comet Wild-2 is located as it sweeps through the solar system at `stardust.jpl.nasa.gov/comets/wildnow.html`, a NASA Web site updated every 10 minutes. And you can track where the Stardust probe is at `stardust.jpl.nasa.gov/mission/scnow.html`, also updated at 10-minute intervals.

For an excellent beginner's guide to meteor observing, forms to submit your meteor counts, and special forms for reporting fireballs, visit the North American Meteor Network at www.namnmeteors.org. Check out Gary Kronk's meteor and comet site at comets.amsmeteors.org and the International Meteor Organization site at www.imo.net.

Spotting sporadic meteors, fireballs, and bolides

When you're outdoors on a dark night and see a "shooting star" (the flash of light from a random, falling meteoroid), what you're probably seeing is a *sporadic* meteor. But if many meteors appear, all seeming to come from the same place among the stars, you're witnessing a *meteor shower*. Meteor showers are among the most enjoyable sights in the heavens; I devote the next section in this chapter to them.

A dazzlingly bright meteor is a *fireball*. Although a fireball has no official definition, many astronomers consider a meteor that looks brighter than Venus to be a fireball. However, Venus may not be visible at the time you see the bright meteor. So how can you decide if you're seeing a fireball?

Here's my rule for identifying fireballs: If people facing the meteor all say "Ooh" and "Ah" (everyone tends to shout when they see a bright meteor), the meteor may be just a bright one. But if people who are *facing the wrong way* see a momentary bright glow in the sky or on the ground around them, it's the real thing. To paraphrase an old Dean Martin tune, when the meteor hits your eye like a big pizza pie, that's a fireball!

Fireballs aren't very rare. If you watch the sky regularly on dark nights for a few hours at a time, you'll probably see a fireball about twice a year. But *daylight fireballs* are very rare. If the sun is up and you see a fireball, mark it down as a lucky sighting. You've seen one tremendously bright fireball. When nonscientists see daytime fireballs, they almost always mistake them for an airplane or missile on fire and about to crash.

Any very bright fireball (approaching the brightness of the half moon or brighter) or any daylight fireball represents a possibility that the meteoroid producing the light will make it to the ground. Freshly fallen meteorites are often of considerable scientific value, and they may be worth good money, too. If you see a fireball that fits this description, write down all the following information so that your account can help scientists find the meteorite and determine where it came from:

1. **Note the time, according to your watch.**

   At the earliest opportunity, check how fast or slow your watch is running against an accurate time source such as a Master Clock at the U.S. Naval Observatory, which you can consult at `tycho.usno.navy.mil/what1.html`.

2. **Record exactly where you are.**

   Chances are good that you don't have a Global Positioning System receiver handy to take a reading of your location, but you can make a little sketch showing where you stood when you saw the fireball — note roads, buildings, big trees, or any other landmarks.

3. **Make a sketch of the sky, showing the track of the fireball with respect to the horizon as you saw it.**

   Even if you're not sure whether you faced southeast or north-northwest, a sketch of your location and the fireball track helps scientists determine the trajectory of the fireball and where the meteoroid may have landed.

After a daylight fireball or a very bright nighttime fireball, interested scientists advertise for eyewitnesses. They collect the information and by comparing the accounts of persons who viewed the fireball from different locations, they can close in on the area where it most likely fell to the ground. Even a brilliant fireball may be only the size of a small stone — one that would fit easily in the palm of your hand — so scientists need to narrow down the search area in order to have a reasonable chance of finding it. If you don't see a call for information after your fireball observation, chances are that the nearest planetarium or natural history museum will accept your report and know where to send it.

A *bolide* is a fireball that explodes or that produces a loud noise even if it doesn't break apart. At least, that's how I define it. Some people use bolide interchangeably with fireball. (You won't find an official agreement on this term; you can find different definitions in even the most authoritative sources.) The noise that you hear is the sonic boom from the meteoroid, which is falling through the air faster than the speed of sound.

When a fireball breaks apart, you see two or more bright meteors at once, very close to each other and heading the same way. The meteoroid that produces the fireball has fragmented, probably from aerodynamic forces, just like an airplane falling out of control from high altitude sometimes breaks apart even though it hasn't exploded.

Often a bright meteor leaves behind a luminous track. The meteor lasts a few seconds or less, but the shining track — or *meteor train* — may persist for many seconds or even minutes. If it lasts long enough, it becomes distorted by the high-altitude winds, just as the skywriting from an airplane above a beach or stadium is gradually deformed by the wind.

You see more meteors after midnight than before because, from midnight to noon, you're on the forward side of the earth, where our planet's plunge through space sweeps up meteoroids. From noon to midnight, you're on the backside, and meteoroids have to catch up in order to enter the atmosphere and become visible. The meteors are like bugs that splatter on your auto windshield. You get many more on the front windshield as you drive down the highway than on the rear windshield, because the front windshield is driving into bugs, and the rear windshield is driving away from bugs.

Watching a radiant sight: Meteor showers

Normally, only a few meteors per hour are visible — more after midnight than before and (for observers in the Northern Hemisphere) more in the fall than in the spring. But on certain occasions every year, you may see 10, 20, or even 50 or more meteors per hour in a dark, moonless sky far from city lights. Such an event is a *meteor shower,* when Earth passes through a great ring of billions of meteoroids that runs all the way around the orbit of the comet that shed them. (I discuss comets in detail later in this chapter.) Figure 4-1 illustrates the occurrence of a meteor shower.

Figure 4-1:
Earth's path
crossing
a belt of
meteoroids
creates a
shower of
meteors.

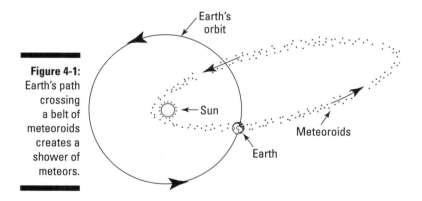

The direction in space or place on the sky where a meteor shower seems to come from is called the *radiant.* The most popular meteor shower is the Perseids, which at its peak produces as many as 80 meteors per hour. (The Perseids get their name because they seem to streak across the sky from the direction of the constellation Perseus, the Hero, their radiant. Meteor showers are usually named for constellations or bright stars [such as Eta Aquarii] near their radiants.)

A few other meteor showers produce as many meteors as the Perseids, but fewer people take the time to observe them. The Perseids come on balmy nights in August, often perfect for skywatching, but the other leading meteor showers — the Geminids and Quadrantids — streak across the sky in the frigid months of December and January, respectively, when the weather is worse and observers' ambitions are limited.

Table 4-1 lists the top annual meteor showers. The dates in the table are the nights when the showers usually reach their peak. Some showers go on for days and others for weeks, raining down meteors at lower rates than the peak values. The Quadrantids may last for just one night or only a few hours.

Table 4-1	Top Annual Meteor Showers	
Shower Name	*Approximate Date*	*Meteor Rate (Per Hour)*
Quadrantids	Jan. 3–4	90
Lyrids	Apr. 21	15
Eta Aquarids	May 4–5	30
Delta Aquarids	July 28–29	25
Perseids	Aug. 12	80
Orionids	Oct. 21	20
Geminids	Dec. 13	100

The Quadrantids' radiant is in the northeast corner of the constellation Bootes, the Herdsman. The meteors are named for a constellation found on 19th-century star charts that astronomers no longer officially recognize. In addition to losing their namesake, the Quadrantids also seem to have lost the comet that spawned them — their origin was a mystery until 2003, when astronomer Petrus Jenniskens found that an object named 2003 EH 1 may be their parent comet.

The Geminids are a meteor shower that seems to be associated with the orbit of an asteroid rather than a comet. However, the "asteroid" is probably a dead comet, which no longer puffs out gas and dust to form a head and tail. The object 2003 EH 1, the likely parent of the Quadrantids, may be a dead comet too. (I discuss comets in the next section.)

The Leonids are an unusual meteor shower that occurs around November 17 every year, usually to no great effect. But every 33 years, many more meteors are present than usual, perhaps for several successive Novembers. Huge numbers of Leonids were seen in November 1966 and again in November 1999, 2000, 2001, and 2002, at least for brief times at some locations. The next great display should come in 2032.

You almost never see as many meteors per hour as I list in Table 4-1. The official meteor rates are defined for exceptional viewing conditions, which few people experience nowadays. But meteor showers vary from year to year, just like rainfall. Sometimes, people do see as many Perseids as listed. On rare occasions, they see many more than expected. Such inconsistency is why keeping accurate records of the meteors that you count can be helpful to the scientific record.

To track meteors, you need a watch, a notebook, a pen or pencil to record your observations, and a dim flashlight to see what you're writing.

The best light for astronomical observations is a red flashlight, which you can purchase or make from an ordinary flashlight by wrapping red transparent plastic around the bulb. Some astronomers paint the lamp with a thin coat of red nail polish. If you use a white light, you dazzle your eyes and make it impossible to see the fainter stars and meteors for 10 to 30 minutes, depending on the circumstances. Letting your vision adjust to the dark is called getting *dark adapted* and is a step you need to take every time you observe the night sky.

The best way to watch and count meteors is to recline on a lounge chair. (You can do pretty well just lying on a blanket with a pillow, but you're more likely to fall asleep in that position and miss the best part of the show.) Tilt your head so that you're looking slightly more than halfway up from the horizon to the zenith (see Figure 4-2) — the optimum direction for counting meteors. And be sure you have a thermos with hot coffee, tea, or cocoa!

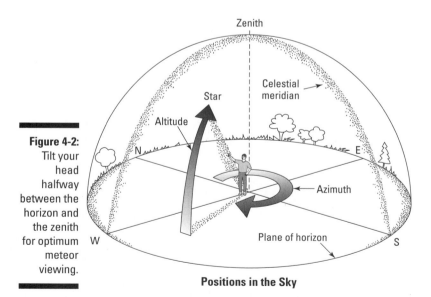

Figure 4-2:
Tilt your head halfway between the horizon and the zenith for optimum meteor viewing.

Positions in the Sky

You don't have to face toward the radiant when you observe a meteor shower, although many people do. The meteors streak all over the sky, and their visible paths may begin and end far from the radiant. But you can visually extrapolate the meteors' paths back in the direction from which they seem to come, and the paths point back to the radiant. That's how you can tell a shower meteor from a sporadic one.

If you do face the radiant, however, you see some meteors that seem to have very short paths, even though they appear fairly bright. The paths appear short because the meteors are coming almost right at you. Fortunately, the shower meteoroids are microscopic and won't make it to the ground.

Photographing meteors and meteor showers

A clear, dark night with no moon is a good time to try to photograph meteors. For the best results, use an old-fashioned, manually operated 35-mm camera — or a modern camera that you can set for totally manual operation. (Use a film camera, not a digital model; digital cameras are unsuited for the long exposures you need to catch meteors.) After you have the proper equipment, follow these steps:

1. **Use an ordinary lens, not a zoom or telephoto lens, and set it for a distance of infinity.**

2. **Set the f/number of the lens at its smallest value.**

   Use a lens that you can set to f/5.6 or smaller — the smaller the better.

3. **Use a film with a speed of ISO 400.**

   Experts often prefer black-and-white film, but color works fine, is easier, and, nowadays, is often cheaper to have developed.

4. **Set your camera on a tripod and point the camera about halfway up the sky or a little higher, facing whichever direction has the least interfering sky glow (light pollution) from city or other lights.**

5. **Set the camera for a time exposure and leave the shutter open for 10 to 15 minutes. Close the shutter, advance the film, and take another exposure.**

If a fireball passes through the part of the sky that the camera faces, however, make a note of the time at once and close the shutter immediately. You have an important picture of a fireball track and you can't add to it by further exposure because the fireball is gone. Begin a new exposure.

6. **When you have the film developed, tell the photo processor to "print all negatives."**

   The operators of photo-processing machines often skip over sky photos, which may appear to be wasted or bad exposures to the non-astronomer.

Photograph a meteor shower the same way that you do a single meteor. But for the best pictures, wait until the radiant (constellation or area from which the meteor shower appears to be coming) is well above the horizon — say 40 degrees or more — before you point your camera toward it. If you capture several shower meteors on the same exposure, their tracks resemble the spokes in a bicycle wheel, all pointing back toward the same place: the radiant.

Here's how to judge altitude above the horizon: The overhead point or zenith is at altitude 90 degrees, and the horizon is at altitude 0 degrees, so halfway from horizon to zenith is 45 degrees altitude, two-thirds the way up is 60 degrees, and so on.

Comets: The Lowdown on Dirty Ice Balls

Comets, great blobs of ice and dust that slowly track across the sky looking like fuzzy balls trailing gassy veils, are popular visitors from the depths of the solar system. They never fail to attract interest. Every 75 to 77 years, the best-known ice ball, Halley's Comet, returns to our neck of the woods. If you missed its appearance in 1986, try again in 2061! If you're impatient, you can see other interesting comets in the meantime. Often a less famous comet, such as Hale-Bopp in 1997, is much brighter than Halley's.

Many people confuse meteors and comets, but you can easily distinguish them after you read the following tips:

- ✔ A meteor lasts for seconds; a comet is visible for days, weeks, or even months.
- ✔ Meteors flash across the sky as they fall overhead, within 100 miles or so of the observer. Comets crawl across the sky at distances of many millions of miles.
- ✔ Meteors are common; comets that you can easily see with the naked eye come less than once a year, on average.

In the following sections, I discuss a comet's structure, famous comets throughout time, and methods you can use to spot a comet.

Making heads and tails of a comet's structure

Historically, astronomers have described comets as having a head and tail or tails. They later named a bright point of light in the head the *nucleus.* Today, we know that the nucleus is the true comet — the so-called dirty ice ball. A comet is a stuck-together mixture of ice, frozen gases (such as the ices of carbon monoxide and carbon dioxide), and solid particles — the dust or "dirt," shown in Figure 4-3. The other features of a comet are just emanations that stem from the nucleus.

Astronomers believe that comets were born in the vicinity of the outer planets, starting near the orbit of Jupiter and extending well beyond Neptune. The comets near Jupiter and Saturn were gradually disturbed by the gravity of those mighty planets and flung far out into space, where they fill a huge, spherical region well beyond Pluto — the Oort Cloud — extending roughly 10,000 A.U. from the sun. (I define the A.U. or Astronomical Unit, a distance equal to about 93 million miles, in Chapter 1.) Other comets were ejected to or were formed and remain in the Kuiper Belt (see Chapter 9), a region that starts around the orbit of Neptune and continues to a distance of about 50 A.U. from the sun, or about 10 A.U. beyond Pluto. Passing stars occasionally disturb these regions and send comets on new orbits, which may take them close to Earth and the sun where we can see them.

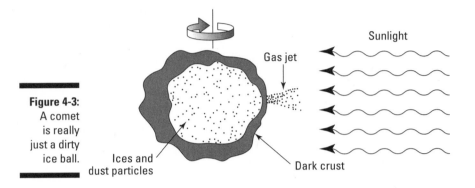

Figure 4-3: A comet is really just a dirty ice ball.

A comet far from the sun is only the nucleus; it has no head or tail. The ice ball may be dozens of miles in diameter or just a mile or two. That may seem pretty small by astronomical standards, and because the nucleus shines only by the reflected light of the sun, a distant comet is very faint and hard to find.

Images of Halley's nucleus, from a European Space Agency probe that passed very close to it in 1986, show that the lumpy, spinning ice ball has a dark crust, like the tartufo dessert (balls of vanilla ice cream, coated with chocolate) served in fancy restaurants. Comets aren't so tasty (I think), but they *are* real treats to the eye. Here and there on Halley's nucleus the probe photographed plumes of gas and dust from geyser-like vents or holes sprayed into space where the sun has barely warmed the surface. Some crust! And in 2004, NASA's Stardust probe got close-up images of the nucleus of Comet Wild-2. This nucleus seems to bear impact craters and is marked with what may be pinnacles made of ice. Those are the cold facts.

As a comet gets closer to the sun, solar heat vaporizes more of the frozen gas and it spews out into space, blowing some dust out, too. The gas and dust form a hazy, shining cloud around the nucleus called the *coma* (a term derived from the Latin for "hair," not the common word for an unconscious state). Almost everyone confuses the coma with the head of the comet, but the head, properly speaking, consists of both the coma and the nucleus.

The glow from a comet's coma is partly the light of the sun, reflected from millions of tiny dust particles, and partly emissions of faint light from atoms and molecules in the coma.

The dust and gas in a comet's coma are subject to disturbing forces that can give rise to a comet's tail(s).

The pressure of sunlight pushes the dust particles in a direction opposite the sun (see Figure 4-4), producing the comet's *dust tail.* The dust tail shines by the reflected light of the sun and has these characteristics:

- A smooth, sometimes gently curved appearance
- A pale yellow color

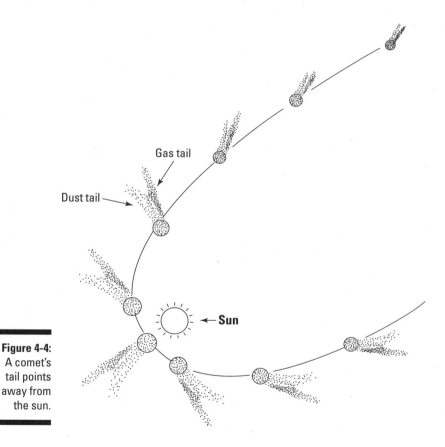

Gas tail

Dust tail

← Sun

Figure 4-4:
A comet's tail points away from the sun.

The other type of comet tail is a *plasma tail* (also called an ion tail or a gas tail). Some of the gas in the coma becomes *ionized,* or electrically charged, when struck by ultraviolet light from the sun. In that state, the gases are subject to the pressure of the *solar wind,* an invisible stream of electrons and protons that pours outward into space from the sun (see Chapter 10). The solar wind pushes the electrified cometary gas out in a direction roughly opposite of the sun, forming the comet's plasma tail. The plasma tail is like a wind sock: It shows astronomers who view the comet from a distance which way the solar wind is blowing at the comet's point in space.

Coma again?

The first rule of viewing is: Get out of town! Although a comet's nucleus may be just 5 or 10 miles in diameter, the coma that forms around it can reach tens of thousands or even hundreds of thousands of miles across. The gases are expanding, like a puff of smoke from a cigarette. As they thin out, they fade and become less visible. So the size of a comet's coma doesn't just depend on how much stuff the comet sheds; it also depends on the sensitivity of the human eye or the photographic film (or electronic detector) that you use to observe it. And the apparent size of the coma also depends on the darkness of the sky in which you view it. A bright comet looks a lot smaller downtown than out in the country where the skies are dark.

In contrast to the dust tail, a comet's plasma tail has

- A stringy, sometimes twisted or even broken appearance
- A blue color

Now and then, a length of plasma tail breaks from the comet and flies off into space. The comet then forms a new plasma tail, much like a lizard that grows a new tail when it loses its first one. The tails of a comet can be millions to hundreds of millions of miles long.

When a comet heads inward toward the sun, its tail or tails stream behind it. When the comet rounds the sun and heads back toward the outer solar system, the tail still points away from the sun, so the comet now follows its tail! The comet behaves to the sun as an old-time courtier did to his emperor: never turning his back on his master. The comet in Figure 4-4 could be going clockwise or counterclockwise, but either way, the tail always points away from the sun.

The coma and tails of a comet are just a vanishing act. The gas and dust shed by the nucleus to form the coma and tails are lost to the comet forever — they just blow away. By the time the comet travels far beyond the orbit of Jupiter, where most comets come from, it consists of only a bare nucleus again. But the dust it loses may produce a meteor shower (which I cover earlier in this chapter) some day, if it crosses Earth's orbit.

Waiting for the "comets of the century"

Every few years, a comet is sufficiently bright enough and in such a good position in the sky that you can easily see it with the naked eye and with small binoculars. I can't tell you when such a comet is coming, because the only comets whose returns astronomers can accurately predict in the near future are small ones that don't get very bright. Nearly all bright, exciting comets are discovered rather than predicted.

Halley's Comet is the only bright comet whose visits astronomers can accurately predict, but it doesn't come around very often. Its appearance in 1910 was widely heralded, and everyone got a look. But an even brighter comet came the same year, the Great Comet of 1910, and no astronomers had predicted its arrival. All you can do is keep looking up. Monitor the astronomy magazines and the Web sites at the end of this section for reports of new comets and then follow the directions to view them. And with luck, you may be the first to spot and report a new comet, in which case the International Astronomical Union will name it after you.

Every five or ten years, a comet comes that is so bright astronomers hail it as "The comet of the century." People have short memories. But stay interested, and you may have a chance to see a fine comet:

- In 1967, Comet Ikeya-Seki was visible in broad daylight next to the sun if you held your thumb up to block the bright solar disk. I'll never forget that sight or my sunburned thumb.

- In 1976, Comet West was visible to the naked eye even in the night sky over downtown Los Angeles, one of the worst places to see celestial objects that I know of.

- In 1983, Comet IRAS-Iraki-Alcock could be seen (with the naked eye) actually moving in the night sky. (Most comets move so slowly across the stars that you may have to wait an hour or more to notice any change in position.)

- In the 1990s, the bright comets Hyakutake and Hale-Bopp appeared out of the blue and were witnessed by millions of people worldwide.

Astronomers, amateur or professional, haven't seen a great comet in the new millennium as of press time, but one is bound to appear, and you may even discover it!

Web sites galore offer information on currently visible comets and photographs of them from amateur and professional astronomers. Most of the time, the current comets are too dim for any but advanced amateur telescopes. Check these three especially good Web sites on a regular basis to make sure you have the latest word:

- The Comet Observation Home Page at NASA's Jet Propulsion Laboratory, at encke.jpl.nasa.gov

- The Current Comets page, with histories and observations, at cometography.com/current_comets.html

- Sky & Telescope's Comet page, which offers hints on how to observe and photograph comets, at skyandtelescope.com/observing/objects/comets/

Hunting for the great comet

Finding a comet isn't difficult, but finding your first one can take years and years. The noted contemporary comet hunter David Levy scanned the sky systematically for nine years before he found his first comet. Since then, he's found over 20 more.

The best telescope to use for comet searching is a *short focus* or *fast* telescope, meaning one whose catalogue specifications include a low f-number (like the f-number of a camera lens) — f/5.6 or, better yet, f/4. And you need to use a low-power eyepiece, such as 20x to 30x (see Chapter 3). The whole idea of the low f/number and the low magnification is to view as large an area of the sky as possible with your telescope. The bright comets that you may be able to discover are few and far between.

A relatively inexpensive telescope to start your comet hunt is the Orion ShortTube 80mm refractor. It has excellent optics at a modest price: $275 when you purchase it with Orion's Paragon tripod. Its f/5.0 focal ratio and 16x magnification are just right for comet hunting. I recommend the optional aluminum carrying case too (about $40), because this telescope is compact enough to carry onboard a plane or fit in a corner of your trunk so you can scan the skies and try to find a comet wherever you go. You can find it on the Orion Telescopes and Binoculars Web site, www.telescope.com. (For more about selecting telescopes, check out Chapter 3.)

You can search for unknown comets in two ways: the easy way and the systematic way. Read on to discover both techniques and for info on reporting a comet.

Locating comets the easy way

The easy way to search for comets is to make no extra effort at all. Just be on the lookout for fuzzy patches when you stare through your binoculars or telescope at stars or other objects in the night sky. Scan the sky for a fuzzy spot (as opposed to stars, which are sharp points of light if your binoculars are in focus). If you pinpoint a fuzzy area, check your star atlas to see whether anything at that location is *supposed* to look fuzzy, such as a nebula or a galaxy. If you find nothing like that on the atlas, you may have found a comet, but before you get too excited, wait a few hours and see if the possible comet moves against the pattern of adjacent stars. If the sun rises or clouds move in the way and block your view, look again on the next night. If the object is indeed a comet, you'll notice that its position has changed with respect to the stars. And, if the fuzz is bright enough, you may be able to spot a tail, which is a dead giveaway that you've spotted a comet.

Locating comets the systematic way

The systematic way to search for comets is based on the precept that you can find them most easily where they're brightest and where the sky is at its darkest. Comets closest to the sun are the brightest, but the sky is darkest in directions far from the direction to the sun.

As a compromise between as far from the sun and as close to the sun as possible, look for comets in the east before dawn over the part of the sky that's

- ✔ At least 40 degrees from the sun (which is below the horizon)
- ✔ No more than 90 degrees from the sun

Remember that there are 360 degrees all the way around the horizon, so 90 degrees is one quarter of the way around the sky.

A desktop planetarium program can help you map out the regions of the constellations that fit this bill for any given night of the year (see Chapter 2 for more about these programs). And, of course, you can look for comets in the west at dusk by following the same two rules about distance from the sun. In my experience, the first few "comets" that you discover will be the contrails from jet airplanes, which catch the rays of the sun at their high altitude, even though the sun has set at your location.

Start at one corner of the sky area that you plan to check and slowly sweep the telescope across the area. Move the telescope slightly up or down and scan the next strip of sky in your search area. You can make every scan from left to right, or you can scan back and forth boustrophedonically (a term from classical times that refers to the plowing of a field with oxen; the oxen plow the first furrow in one direction and then come back across the field, plowing in the opposite direction).

Playing the space object name game

If you discover a comet, the International Astronomical Union will name it after you and possibly also after the next one or two people who independently report it.

If you discover a meteor, you don't have time to name it before it disappears. You can try shouting "John," but the name won't catch on and you may attract undue attention. The only meteors that get named are the spectacular ones seen by thousands of people. They get names such as the "Great Daylight Fireball of August 10, 1972," but no official procedure governs how this happens.

If you discover a meteorite, it's named for the town or other local area where you pick it up. The meteorite belongs to the owner of the land where you find it, and if you find it on U.S. government land, such as a national park or forest, it goes to the Smithsonian Institution.

If you discover an asteroid, you can recommend a name for the asteroid, but it can't be yours. (See Chapter 7 for more about asteroids.)

You'll have an easier time impressing your friends by telling them of your boustrophedonic comet search project than in actually discovering a comet. It will give your ego a boost (unless your friends decide that you're plowed).

Reporting a comet

When you discover a comet, follow the directions on the Web site of the International Astronomical Union's Central Bureau for Astronomical Telegrams (which doesn't use telegrams any more) and report it by e-mail. The site is `cfa-www.harvard.edu/iau/cbat.html`.

The Bureau doesn't appreciate false alarms, so try to get a stargazing friend to check out your discovery before you spread the word. If the find checks out, you — as the amateur discoverer of a comet — may be eligible for a cash share of the Edgar Wilson Award, which is described on the Central Bureau Web site (`cfa-www.harvard.edu/iau/special/EdgarWilson.html`).

But even if you never discover a comet, and most astronomers never do, you can enjoy comets that others discover.

Artificial Satellites: Enduring a Love-Hate Relationship

An artificial satellite is something that people build and launch into space, where it orbits the earth or another celestial body. The earth-orbiting artificial satellites show us the weather, monitor El Niño, relay network television programs, and stand guard against intercontinental missile launches by hostile powers. And they can also be used for astronomy.

The Hubble Space Telescope is an artificial satellite, and astronomers love it. It gives us unparalleled views of the stars and distant galaxies and lets us view the universe in ultraviolet and infrared light that's otherwise blocked by the thick layers of Earth's atmosphere.

But artificial satellites can also catch the rays of the setting sun or even the sun that has already set for observers at ground level. Capturing the light of the sun, they represent points of light that may move across the part of the sky where an astronomer is making a time exposure photograph of faint stars. Astronomers don't appreciate this interference. Worse yet, some artificial satellites broadcast at radio frequencies that interfere with the "big dish" and other radio antennas that astronomers use to receive the natural radio emanations from space. The celestial radio waves may have traveled for five billion years from a quasar, or they may have taken 5,000 years to reach us from another solar system in the Milky Way, possibly bearing a greeting from benevolent aliens who want to send us the cure for cancer. But just as the radio waves arrive, a blaring tone and strident modulations from a satellite passing over the observatory interfere with our reception. We may never know what the news is from the Alpha Centauri system.

So astronomers love satellites when they do something good for us and hate them when they interfere with our observations. But to make the best of a bad thing, amateur astronomers have become enthusiastic viewers and photographers of artificial satellites passing overhead.

Skywatching for artificial satellites

Hundreds of operating satellites are orbiting Earth, along with thousands of pieces of orbiting space junk — nonfunctional satellites, upper stages from satellite launch rockets, pieces of broken and even exploded satellites, and tiny paint flakes from satellites and rockets. On the ground, the space shuttle is a manned rocket, but in space, it orbits Earth as a big artificial satellite.

You may be able to glimpse the reflected light from any of the larger satellites and space junk, and powerful defense radar can track even very small pieces.

The best way to begin observing artificial satellites is to look for the big ones — such as NASA's International Space Station, the Hubble Space Telescope, and a space shuttle (when in space on a mission) — and the bright flashing ones (the dozens of Iridium communication satellites).

Looking for a big or bright artificial satellite can be reassuring to the beginning astronomer. Predictions of comets and meteor showers are often mistaken, the comets always seem to be fainter than you expect, and often you see fewer meteors than advertised. But artificial satellite viewing forecasts are usually right on. You can amaze your friends by taking them outside on a clear early evening, glancing at your watch, and saying "Ho hum, the International Space Station should be coming over about there (point in the right direction as you say this) in just a minute or two." And it will!

Want to know what to watch for? I've got you covered. Here are some characteristics you can pinpoint for both large and bright satellites:

✔ A satellite such as the Hubble Space Telescope or the International Space Station generally appears in the evening as a point of light, moving steadily and noticeably from west to east in the western half of the sky. It moves much too slowly for you to mistake it for a meteor, and it moves much too fast to resemble a comet. You can see it easily with the naked eye, so it can't be an asteroid, and anyway, it moves much faster than an asteroid.

Sometimes, you may confuse a high-altitude jet plane with a satellite. But take a look through your binoculars. If the object in view is an airplane, you should be able to distinguish running lights or even the silhouette of the plane against the dim illumination of the night sky. And when your location is quiet, you may be able to hear the plane. You can't hear a satellite.

✔ An Iridium satellite is a wholly different viewing situation: It usually appears as a moving streak of light that gets remarkably bright and then fades out after several seconds. It moves much slower than a meteor. And an Iridium flare or flash is often brighter than Venus, second in brilliance only to the moon in the night sky. The sun, located below your horizon, reflects off one of the door-sized, flat, aluminum antennas on the satellite to cause the flash of light. At star parties, people cheer when they spot an Iridium flare, just like when folks see a fireball. You can even see some Iridium flares in daylight.

And consider this: More than 60 Iridium satellites are in orbit. They interfere with astronomy, and astronomers want them to disappear, but at least the satellites have a "flare" for entertaining us.

Finding satellite viewing predictions

Some newspapers and television weather persons give daily or occasional forecasts for viewing satellites from your local area. You can get more detailed information whenever you want it by consulting these Web sites:

✔ For the International Space Station, *Sky & Telescope* offers observing predictions on its Almanac page at skyandtelescope.com/observing/almanac. Change the default location (Greenwich, England) to your city and put in the date and time when you want to start looking for the Space Station.

✔ For Iridium communication satellites, the most convenient forecasts are available from the Heavens-Above Web site at www.heavens-above.com. Enter your location and time and Heavens-Above computes away for a minute or two and then gives you a table of upcoming viewing opportunities.

✔ Heavens-Above is also the place for Hubble Space Telescope viewing info. Click on the Hubble link in the section on "Satellites" to get a viewing schedule for your location.

After you succeed in viewing some bright artificial satellites, you can try photographing them. Follow the directions in the sidebar "Photographing meteors and meteor showers" earlier in this chapter. All you need is a camera suitable for taking time exposures, a steady tripod, and some fast film.

Part II
Going Once Around the Solar System

The 5th Wave   By Rich Tennant

In this part . . .

Guess what? Men aren't from Mars, and women aren't from Venus. In fact, neither planet can support life as we know it. Venus is too hot, Mars is too cold, and neither has any liquid water that scientists know of.

This part explains what the planets of our solar system are really like. Did Mars ever support life? What about Jupiter's moon, Europa? I tell you what scientists know as of now.

And if you've watched one of those "Oh no, a big asteroid is headed for Earth!" movies, you may be wondering whether to take a threat like that seriously. I include a chapter that explains asteroids and tells you the truth about the risk of their striking the earth.

Chapter 5

A Matched Pair: Earth and Its Moon

In This Chapter

▶ Seeing Earth as a planet

▶ Understanding Earth's time, seasons, and age

▶ Focusing on the moon's phases and features

*P*eople often think of planets as objects up in the sky, like Jupiter and Mars. The ancient Greeks — and people for centuries thereafter — made a distinction between Earth, which they regarded as the center of the universe, and the planets. They thought of the planets as little lights in the sky that revolved around Earth.

Today we know better. Earth isn't the center of the universe. It isn't even the center of our solar system; the sun owns that title. The moon orbits around Earth, along with hundreds of artificial satellites (see Chapter 4), and that's about it. And joining Earth in orbit around the sun are eight other planets in the solar system, a number of other moons, a belt of asteroids, millions of comets, and more. Nevertheless, as far as we know now, life exists in our solar system only on Earth.

Earth has fallen from its exalted place in human thought as the center of the universe to its true, but still significant, status: our home planet. And you can't find another place in the solar system that's quite like home.

Earth is what astronomers call a *terrestrial* planet — a kind of circular definition, because terrestrial means "earthly." But the scientific meaning is a planet made of rock that orbits the sun. The four planets closest to the sun are our solar system's terrestrial planets: Mercury, Venus, Earth, and Mars, in order of distance from the sun.

Some people consider Earth's moon a terrestrial planet and regard the Earth-Moon system as a double planet. For aliens seeking to visit us, that distinction probably helps: "Just head for that yellow-white star in Sector 49,832 of the Orion Arm, in the Milky Way, and home in on the third rock from that sun; it's a double planet and easy to spot."

Putting Earth Under the Astronomical Microscope

Earth is unique among the known planets. In the following sections, I tell why, briefly summarizing some of its main characteristics and how they play into astronomical topics like time and the seasons. And in case you forgot what it looks like, you can check out a nice NASA photo in the color section, which shows the earth and moon together.

One of a kind: Earth's unique characteristics

What's so special about Earth? For starters, we inhabit the only planet we know of with

- **Liquid water at the surface.** Earth has lakes, rivers, and oceans, unlike any other known planet. Unfortunately, it has tsunamis and hurricanes too. The oceans cover 70 percent of the surface of Earth.

- **Plentiful amounts of oxygen in the air.** The air on Earth contains 21 percent oxygen; no other planet has more than a trace of oxygen in its present atmosphere, as far as we know.

- **Plate tectonics (also known as *continental drift*).** Earth's crust is composed of huge moving plates of rock; where plates collide, earthquakes occur, and new mountains rise. New crust emerges at the mid-ocean ridges, deep beneath the sea, causing the seafloor to spread. (To find out about an interesting seafloor property, see the sidebar "Earth's seafloor and its magnetic properties" later in this chapter.)

- **Active volcanoes.** Hot molten rock, welling up from deep beneath the surface, forms huge volcanic landforms such as the Hawaiian Islands. Volcanoes erupt somewhere on Earth every day.

- **Life, intelligent or otherwise.** I'll let you be the judge on intelligence, but from one-celled amoebas, bacteria, and viruses to flowers and trees, fish and fowl, and insects and mammals, Earth has life in abundance.

  Researchers are investigating tantalizing indications that Mars and Venus may once have shared some of these traits with Earth (see Chapter 6). But as far as we know, they don't have life now, and we don't have proof that they ever did.

Scientists believe that the presence of liquid water on the surface of Earth is one of the main reasons why life flourishes here. You can easily imagine advanced life forms on other worlds. You see them on television and in the movies. But the images you see are all imaginary. Scientists don't have convincing evidence for any life, past or present, anywhere but on Earth.

Enjoying the northern lights

The aurora is one of the most beautiful sights of the night sky, and for many people, a rare one. Depending on whether you live in the Northern Hemisphere or the Southern Hemisphere, you can see the *aurora borealis* (northern lights) or the *aurora australis* (southern lights), respectively.

Auroras appear when streams of electrons from Earth's magnetosphere rain down on the atmosphere below, stimulating oxygen and other atoms to shine. The eerie glow in the dark night sky may remain stationary for minutes to hours or constantly change (making it hard for a beginning observer to identify). It can shimmer, pulsate, or even flash around the sky. The aurora may appear to you in many forms; here are a few of the most common:

✔ **Glow:** The simplest form of auroral display. The glow resembles a part of the sky where a thin cloud reflects moonlight or city lights. But you don't see any clouds — just the eerie light of an aurora.

✔ **Arc:** Shaped like a rainbow but with no sunlight to produce one. A steady or pulsating green arc is the most common type of arc, but sometimes faint red arcs appear.

✔ **Curtain:** Also called drapery. This spectacular auroral form resembles a billowing curtain at a theater, where nature is the star of the show.

✔ **Rays:** One or more long, thin bright lines in the sky, appearing like faint beams from the heavens.

✔ **Corona:** High overhead, a crown in the sky, with rays emanating in every direction.

Auroras occur constantly in two geographical bands around Earth at high northern and southern latitudes. Folks who live beneath these two *auroral ovals* see auroras every night. But you may encounter big exceptions: When a great disturbance in the solar wind strikes the magnetosphere, the ovals move toward the equator. People in the *auroral zones* (the lands beneath the ovals) may miss their aurora, but skygazers toward the equator who rarely see them are treated to a great show. The most likely times to see bright auroras outside the auroral zones are the first few years after the peak of the sunspot cycle, so keep your eyes open for auroras around 2013 and the following few years. If you don't want to wait that long for the aurora to come to you, visit Alaska or Norway, where you're near the northern auroral oval and can see the northern lights on most any clear night.

Check out the daily appearance of the auroral ovals with views and data from NASA, NOAA (the National Oceanographic and Atmospheric Administration), and U.S. Air Force satellites at the Solar Terrestrial Dispatch site, www.spacewe.com. You can find forecasts of auroral activity, and you can report your aurora sightings on the Auroral Activity Submission Form.

Spheres of influence: Earth's distinct regions

Figure 5-1 shows four views of Earth as seen from space. Earth's patterns of land, sea, and clouds are clearly visible.

Scientists classify the regions of Earth into

- The *lithosphere:* The rocky regions of our planet
- The *hydrosphere:* The water in the oceans, lakes, and elsewhere on Earth
- The *cryosphere:* The frozen regions — notably the Antarctic and Greenland ice caps
- The *atmosphere:* The air from ground level on up for hundreds of miles
- The *biosphere:* All the living things on Earth — on land, in the air and water, and underground

So you're part of the biosphere that lives on the lithosphere, drinks from the hydrosphere, and breathes the atmosphere. (You can also take a tour to the cryosphere.) I don't know anywhere else in space where you can do all that.

Figure 5-1:
Four views showing the changing face of Earth.

Courtesy of NASA

In addition to the regions I describe in the previous list, one more important part of our planet is the *magnetosphere,* which plays an important part in protecting Earth from many of the dangerous emanations from the sun (see Chapter 10). Sometimes called Earth's radiation belts (or the Van Allen radiation belts, named for James Van Allen, a U.S. physicist who discovered them with America's first artificial satellite, Explorer 1), the magnetosphere consists of electrically charged particles — mostly electrons and protons — that bounce back and forth above Earth, trapped in its magnetic field.

Occasionally, some of the electrons escape and rain down on Earth's atmosphere below, striking atoms and molecules and making them glow. That glow is the aurora (see the previous section, and check out the sidebar "Enjoying the northern lights," earlier in this chapter, for more about viewing auroras).

The solid surface of Earth — the part you stand on — is the crust. Beneath the crust are the mantle and the core. The core is largely iron and nickel and very hot, reaching about 12,000°F (about 7,000°C) at the center. And the core is layered, too: The outer core is in a molten state, and the inner core is solid.

The extremely high pressure of the overlying layers makes the hot iron in the inner core solidify. As Earth cools down over millions of years in the future, the solid part at the center will increase in size at the expense of the surrounding molten core, like an ice cube growing as the surrounding liquid cools.

Earth's core is far below our digging range, but it produces an effect that anyone can observe at the surface. Moving streams of molten iron in the outer core generate a magnetic field that reaches out through the whole planet and far into space, called the *geomagnetic field.*

The geomagnetic field

- ✔ Makes a compass needle point
- ✔ Provides an invisible guidance system for homing pigeons, some migratory birds, and even some ocean-dwelling bacteria
- ✔ Forms the magnetosphere far above Earth
- ✔ Shields Earth from incoming electrically charged particles from space, such as the solar wind and many cosmic rays (high speed, high energy particles that come from explosions on the sun and from distant points in space)

The geomagnetic field is a global planetary magnetic field, meaning it extends above all parts of Earth and is continuously being generated. Mars, Venus, and our moon all lack a global magnetic field like Earth's, and this key difference gives scientists information about the cores of those objects. For more on the lunar core, see the section "Quite an impact: A theory about the moon's origin" later in this chapter.

Earth's seafloor and its magnetic properties

According to geophysical surveys, patterns of magnetized rock exist in the seafloor on either side of mid-ocean ridges. The rock became magnetized as it cooled from the molten state, trapping and "freezing in" some of Earth's magnetic field that pervaded it as the rock solidified. So the seafloor rock resembles a magnet, with a magnetic field that has strength and direction. After the rock solidified, its magnetic field could change no longer, and it became a fossil magnetic field. It's like a fossil dinosaur that remains forever in the shape it had at death.

The patterns discovered near the mid-ocean ridges consist of stripes of magnetized rock, hundreds of miles long, that parallel the ridges and alternate in polarity. One stripe has a north magnetic polarity, like the end of a bar magnet that attracts a north-seeking compass needle, and the next stripe has the opposite polarity, and so on.

The alternating stripes of oppositely magnetized rock are due to the new rock emerging from the mid-ocean ridges, cooling and magnetizing, and moving away from the ridges as even newer rock pushes it along. The oppositely magnetized stripes show that the geomagnetic field itself periodically reverses direction, like a bar magnet that you turn 180 degrees at intervals — except that the intervals for the geomagnetic field are probably hundreds of thousands of years.

An unknown process causes the geomagnetic field, generated deep in Earth's core, to reverse every so often. That effect is preserved in the fossil magnetic fields of the rock at the seafloor and in rock on the continents that previously lay beneath the sea.

Why mention all this seafloor stuff in a book on astronomy? Because this unique property of Earth may correspond to a phenomenon discovered on Mars. As scientists consider the evidence gathered on the various terrestrial planets, including Earth, we find similarities and differences that help us understand them better. Such research is called *comparative planetology,* which I cover in more detail in the descriptions of Mars and Venus in Chapter 6.

Examining Earth's Time, Seasons, and Age

It may be hard to believe because you can't walk 5 feet anymore without having access to a clock, but the rotation of Earth was the original basis of our system of measuring time, and we now know that the orbital motion of Earth and the tilt of its axis produce the seasons. Our seasonal ring-around-the-rosy (or yellowy, in this case) sun dance has been going on for a long time; Earth is about 4.6 billion years old.

Orbiting for all time

Nowadays, scientists have atomic clocks that measure time with great precision. But originally and until modern times, our system of time was based on the rotation of the earth.

Knowing how time flies

Earth rotates once on its axis in 24 hours. It turns from west to east (or counterclockwise as visible from above the North Pole), and it orbits the sun counterclockwise (as visible from space way above the North Pole). The length of the day, 24 hours, is the average time it takes for the sun to rise and set and rise again. This process is called *mean solar time* and is equivalent to the standard time on your watch.

The length of the day, therefore, is 24 hours of mean solar time. And a year consists of approximately 365 days, the time that it takes Earth to make one complete orbit around the sun.

Because Earth moves around the sun, the time when you see the sun rise depends on both the rotation of Earth and Earth's orbital motion.

Earth turns once in 23 hours, 56 minutes, and 4 seconds with respect to the stars. That amount of time is called the *sidereal day.* (Sidereal means pertaining to the stars.) Notice that the difference between 24 hours and 23 hours, 56 minutes, and 4 seconds is 3 minutes and 56 seconds, which is just about $\frac{1}{365}$ of a day. The difference is no coincidence: It happens because during a day, Earth moves through $\frac{1}{365}$ of its orbit around the sun.

Astronomers used to depend on special clocks called sidereal clocks, which measured sidereal time by registering 24 sidereal hours during an interval of 23 hours, 56 minutes, and 4 seconds of mean solar time. The sidereal hours, minutes, and seconds are all slightly shorter than the corresponding units of solar time. Using sidereal clocks enabled astronomers to keep track of the stars in order to point telescopes correctly. But we don't need to do that anymore. Computer programs that point telescopes or that picture the sky on a desktop planetarium, as I describe in Chapter 2, do the mathematics for you, so you can simply use the standard time at your location to figure out where different stars and constellations appear in the sky.

On the other hand, astronomers still adhere to the custom of reporting astronomical observations on a common system called *Universal Time (UT)* or *Greenwich mean time.* UT is simply the standard time at Greenwich, England. If you live in North America, the standard time at your location is always earlier than the time in Greenwich. For example, in New York City, the sun rises about 5 hours after it rises in Greenwich. When the clock strikes 6 a.m. in Greenwich, clocks in New York are turning to 1 a.m.

A more precisely defined time, *Coordinated Universal Time* or *UTC,* which is identical to UT for all practical purposes, is the official international standard.

Finding the right time

In the United States, the U.S. Naval Observatory (USNO) in Washington, D.C. is in charge of time. You can get the UTC any time you want it from the USNO Web page at `tycho.usno.navy.mil/what.html`. The USNO site also has a place where you can find the Local Apparent Sidereal Time at your location. The *Local Apparent Sidereal Time* equals the right ascension (see Chapter 1) of the stars on your meridian — the imaginary line from the zenith to the south point on the horizon. A star is best placed for observation when it's on the meridian.

To determine the standard time zone that applies at just about any other place in the world, and to convert it to Universal Time, consult the World Time Zone Map of Her Majesty's Nautical Almanac Office at `aa.usno.navy.mil/AA/faq/docs/world_tzones.html`.

Generally speaking, daylight saving time (called Summer Time in the U.K.) is an hour later than standard time at the same geographic location. But not all places observe daylight saving time. For example, Arizona, which gets plenty of sunshine year-round, never goes on daylight saving time.

Tilting toward the seasons

Teaching students about the cause of the seasons is about the most frustrating task of any astronomy professor. No matter how carefully the professor explains that the seasons have nothing to do with how far we are from the sun, many students don't grasp it. Surveys taken, even at Harvard University graduation, show that bright college graduates think that summer is when Earth is closest to the sun and winter is when Earth is farthest from the sun.

What students forget is that when summer comes in the Northern Hemisphere, the south experiences winter. And when Australians are surfing in the summer, people in the United States are wearing their winter coats. But Australia and the United States are on the same planet. Earth can't be farthest from the sun and closest to the sun at the same time. Earth's a planet, not a magician.

The actual cause of the seasons is the tilt of Earth's axis (see Figure 5-2). The *axis,* the line through the north and south poles, isn't perpendicular to the plane of Earth's orbit around the sun. Actually, the axis is tilted by 23½ degrees from the perpendicular to the orbital plane. The axis points north to a place among the stars — in fact, near the North Star (at least in the short term; the axis slowly changes its pointing direction, so the North Star in one era will no longer be the North Star in the far future).

At present, the North Star, also called Polaris, is the star Alpha Ursae Minoris, located in the Little Dipper asterism of the Little Bear constellation, Ursa Minor. If you get lost at night and want to "bear" north, set your sights on the Little Dipper (see Chapter 3 for more about finding Polaris).

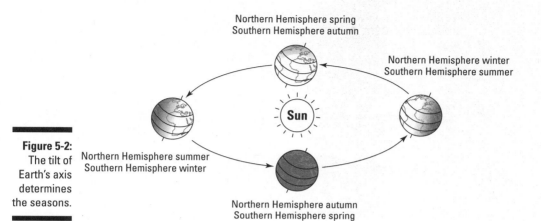

Figure 5-2:
The tilt of
Earth's axis
determines
the seasons.

Northern Hemisphere spring
Southern Hemisphere autumn

Northern Hemisphere winter
Southern Hemisphere summer

Sun

Northern Hemisphere summer
Southern Hemisphere winter

Northern Hemisphere autumn
Southern Hemisphere spring

The axis of Earth points "up" through the North Pole and "down" through the South Pole. When Earth is on one side of its orbit, the axis pointing up also points roughly toward the sun so that the sun looms high in the sky at noon in the Northern Hemisphere. Six months later, the axis points up and roughly away from the sun. Actually, the axis always points in the same direction in space, but Earth has now moved to the opposite side of the sun.

Summer occurs in the Northern Hemisphere when the axis pointing up through the North Pole points roughly at the sun. When that happens, the sun at noon is higher in the sky than at other seasons of the year, so it shines more directly on the Northern Hemisphere and provides more heat. At the same time, the axis pointing down through the South Pole points away from the sun, so it shines lower in the sky at noon than at any other season of the year, creating less direct sunlight — and thus, winter in Australia.

We enjoy more hours of sunlight in the summer because the sun is higher in the sky. It takes longer for the sun to rise to that height, and it takes longer to set.

As we orbit the sun, it seems to move through the sky, following a circle called the *ecliptic,* which I mention in Chapter 3. The ecliptic is tilted with respect to the equator by exactly the same angle as Earth's tilt on its axis: 23½ degrees. Here are some key events in the sun's annual journey around the ecliptic:

- When the sun crosses from "below" (south) the equator to "above" (north), we experience the first day of spring, called the *vernal equinox.*

- When it reaches the farthest point north on the ecliptic, we have the *summer solstice.*

- When it crosses the equator going back down south, fall begins with the *autumnal equinox.*

- And when it gets as far south as possible on the ecliptic, we have the *winter solstice.*

In the Northern Hemisphere, the summer solstice is the day with the most hours of sunlight during the year, because the sun attains its highest position in the sky — taking the longest time to reach that height and come back down to the horizon again. By the same token, the winter solstice in the Northern Hemisphere is the day with the shortest amount of daylight during the year.

And that's the long and the short of time and seasons.

Estimating Earth's age

Measuring radioactivity is the only accurate way we have to date very old things on Earth or in the solar system. Some elements, such as uranium, have unstable forms called *radioactive isotopes*. A radioactive isotope turns into another isotope of the same element, or into a different element, at a rate determined by the *half-life* of the radioactive substance. If the half-life is 1 million years, for example, half of the radioactive isotope that was originally present will have turned into another substance (called the *daughter isotope*) by the time 1 million years have elapsed, leaving half still radioactive. And half of the remaining half turn into daughter isotope atoms in another million years. So after 2 million years, only 25 percent of the original radioactive isotope atoms still exist. After 3 million years, only 12½ percent remain. And so on.

When the original radioactive isotope atoms, called the *parent atoms,* and the daughter atoms are trapped together in a piece of rock or metal, such as a meteorite, scientists can count the atoms' respective numbers to determine how old the rock is in a process called *radioactive dating.*

Scientists have used radioactive dating to determine that the oldest rocks on Earth are about 3.8 billion years old. However, Earth is undoubtedly much older than that. Erosion, mountain building, and *volcanism* (the eruption of molten rock from within the earth, including the formation of new volcanoes) constantly destroy the rocks at the surface, so the original surface rocks of Earth are long gone.

Meteorites, however, yield radioactive dates as old as 4.6 billion years. Meteorites are considered debris from asteroids, and asteroids are thought to be debris from the very early solar system, when the planets first formed (see Chapter 7 for more about asteroids).

So scientists think that Earth and other planets are about 4.6 billion years old. Earth's moon, however, is a little younger, as I explain in the next section.

Making Sense of the Moon

The moon is 2,160 miles (3,476 kilometers) in diameter, slightly more than ¼ the diameter of Earth. The moon has no meaningful atmosphere, just a trace of hydrogen, helium, neon, and argon atoms, along with other traces in even lesser quantities. It's made of solid rock (see Figure 5-3). Its mass is only ⅟₈₁ the mass of Earth, and its density is about 3.3 times the density of water, which is noticeably less than the density of Earth (5.5 times the density of water).

The following sections give you the lowdown on the moon's phases, lunar eclipses, and its geology (including handy tips for viewing a variety of lunar features). I also share a theory about the moon's origin.

Get ready to howl: Phases of the moon

Except during a lunar eclipse (see the next section), half the moon is always in sunlight and half is always in night. But these light and dark hemispheres don't, contrary to popular belief, correspond to the lunar near side and the lunar far side. Those sides are the hemispheres that point toward and away from Earth, which are always the same. The lunar halves in sunlight and in night are the hemispheres that face toward and away from the sun. And they always change as the moon moves around Earth (see Figure 5-4).

Figure 5-3:
The moon is made of rocks and rilles, craters, and dried lava plains — not a speck of cheese in sight.

Courtesy of NASA

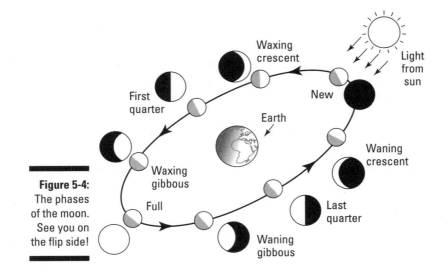

Figure 5-4:
The phases
of the moon.
See you on
the flip side!

New Moon is the beginning of the monthly lunar cycle, or *lunation*. At this time, the near side faces away from the sun, making it the dark side. A few hours or days later, the moon is a new crescent, or *waxing crescent,* meaning a crescent moon whose bright area is getting larger. This phase happens as the moon moves away from the sun-earth line while orbiting Earth. Fully half of the moon is always lit up, facing the sun, but during a crescent moon, we can't see most of this illuminated area that faces away from Earth.

As the moon moves around its orbit, it reaches a point where the earth-moon line is at right angles to the earth-sun line. At this stage, we see a *half Moon,* which astronomers call a *quarter Moon.*

How can a half equal a quarter? It can't if you're trying to make change, but an astronomer can easily make it work. Half of the lunar near side — the part facing Earth — is lit up, so people call it a half Moon. But the illuminated portion of the moon that we see is only half the bright hemisphere that faces the sun, and half of a half is a quarter. Bet your friends that a quarter can be a half. You'll win, and you can pocket the change.

When the illuminated part of the moon that we can see grows larger than the quarter (half) Moon and smaller than the full Moon, astronomers call it a *waxing gibbous Moon.*

When the moon is on the far side of its orbit, opposite the sun in the sky, the lunar hemisphere that faces Earth is fully lit, creating a *full Moon.* As the moon continues around its orbit, the illuminated portion gets smaller and the moon becomes gibbous again, less than full and more than a quarter Moon (a *waning gibbous Moon*). Soon the moon appears as a quarter Moon again, called *last quarter.* As the moon nears the line between Earth and the sun, it becomes a *waning crescent Moon*. Soon it becomes a new Moon again, and the cycle of phases starts over.

People often ask why an eclipse of the sun doesn't occur every month at new Moon. The reason is that Earth, the moon, and the sun usually aren't all exactly on a line at a new Moon. When they are, an eclipse of the sun results. When the three bodies are all on a line at a full Moon, we witness an eclipse of the moon.

Earth has phases, too, just like the moon! To see them, however, you need to head into space and look back at Earth from a distance. When folks on Earth see a beautiful full Moon, an observer standing on the moon would enjoy a "new Earth," and when earthlings experience a new Moon, viewers on the moon see a "full Earth."

In the shadows: Watching lunar eclipses

A lunar eclipse occurs when a full Moon is exactly on the line from the sun to the earth. The moon is then in Earth's shadow, or the *umbra*. You can safely look at a lunar eclipse, as long as you don't bump into something in the dark or stand in the road.

During a total eclipse of the moon, you can still see the moon, although it's immersed in Earth's shadow (see Figure 5-5). No direct sunlight falls on it, but some light from the sun gets bent around the edges of Earth's atmosphere (as visible from the moon) and falls on the moon. The sunlight gets strongly filtered as it passes through our atmosphere, so mostly the red and orange light gets through. This effect differs from one lunar eclipse to the next, depending on meteorological conditions and the clouds in Earth's atmosphere. The totally eclipsed moon, therefore, can look a dull orange, an even duller red, or a very dark red. Sometimes, you can barely make out the eclipsed moon at all.

The dates for upcoming total lunar eclipses through the year 2015 are

March 3, 2007

August 28, 2007

February 21, 2008

December 21, 2010

June 15, 2011

December 10, 2011

April 15, 2014

October 8, 2014

April 4, 2015

September 28, 2015

To prepare in advance for the upcoming eclipses, you can find plenty of information on the exact times and on the part of Earth where the eclipse will be visible. Take a look in *Night Sky, Astronomy,* and *Sky & Telescope* magazines and on their Web sites (www.nightskymag.com, www.astronomy.com, and skyandtelescope.com) as the dates of the eclipses approach.

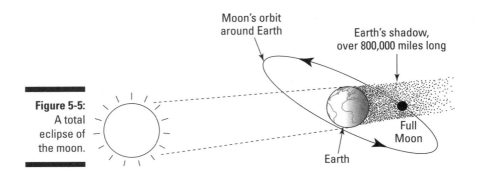

Figure 5-5: A total eclipse of the moon.

Total eclipses of the moon are as common as total eclipses of the sun, but you see them more often at any given place, because a total eclipse of the sun is visible only along a narrow band on Earth called the *path of totality.* But when Earth's shadow falls on the moon, you can see the eclipsed moon from all over half of Earth, wherever night has fallen.

Partial eclipses aren't quite as interesting. During a *partial eclipse,* only some of the full Moon falls within Earth's shadow. The moon just appears to be at a different phase. If you don't know that an eclipse is in progress — or that the full Moon phase is underway — you aren't aware that a unique astronomical event is occurring. You may simply pass it off as a quarter or crescent moon. But if you keep looking for an hour or so, you can see the full Moon come out from Earth's shadow.

Hard rock: Viewing the moon's geology

The entire moon is pockmarked with craters of every size, from microscopic pits to basins hundreds of miles in diameter. The largest is the South Pole–Aitken Basin, which is about 1,600 miles (2,600 kilometers) across. Objects (asteroids, meteoroids, and comets) that struck the moon — very long ago, for the most part — caused these craters. The microscopic craters, which scientists have found on rocks brought back by astronauts from the surface of the moon, are caused by micrometeorites — tiny rock particles flying through space. All the craters and basins are known collectively as *impact craters* to distinguish them from volcanic craters.

The moon has experienced volcanism, but it took a form different from Earth's. The moon has no *volcanoes,* or large volcanic mountains with craters at the top. But it does have small volcanic domes, or round-topped hills like those that occur in some volcanic regions on Earth. In addition, sinuous channels on the lunar surface (called *rilles*) appear to be lava tubes, also a common landform in volcanic areas on Earth (such as Lava Beds National Monument in northern California). Most notably, the moon has huge lava plains that fill the bottoms of the large impact basins. These lava plains are called *maria,* the Latin word for seas. (When you look up and see the Man in the Moon, the dark areas that make up some of his features are the maria.)

Some early scientists thought that the maria could be oceans. But if they were oceans, you could see bright reflections of the sun from them, just as you do when you look down at the sea from an airplane during the day. The larger, bright areas in the Man in the Moon are the *lunar highlands,* which are heavily cratered areas. The maria have craters, too, but fewer craters per square mile than the highlands, which means that the maria are younger. Huge impacts created the basins where the maria are located. These impacts obliterated preexisting craters. Later, the basins filled with lava from below, wiping out any new craters that had formed after the huge impacts. All the craters you can see in the maria now are from impacts that happened after the lava froze.

In the late 1990s, a NASA spacecraft called the Lunar Prospector obtained indirect evidence indicating that there may be frozen water in the bottoms of a few craters near the North and South Poles of the moon, where the sun never shines. The area includes the South Pole–Aitken Basin, a likely target for future space missions. The sun, at best, is low on the horizon near the poles of the moon; the crater rims block the sun from shining on parts of the crater bottoms. The ice may have come from comets that struck the moon long ago, because comets are largely ice and occasionally impact celestial bodies. But evidence suggests that no other water is present on the moon.

Ready to observe the moon's near side and find out the scoop on the moon's far side? Check out the following sections.

Observing the near side

The moon is one of the most rewarding objects to observe. You can see it when the sky is hazy or partly cloudy, and at times it's visible during the day. You can see craters with even the smallest telescopes. And with a high-quality small telescope, you can enjoy hundreds, and maybe thousands, of lunar features, including impact craters, maria, lunar highlands, rilles, and other features, including

 ✔ **Central peaks:** Mountains of rubble thrown up in the rebound of the lunar surface from the effects of a powerful impact. Central peaks are found in some, but not all, impact craters.

✔ **Lunar mountains:** The rims of large craters or impact basins, which may have been partly destroyed by subsequent impacts, leaving parts of their walls standing alone like a range of mountains, although not the type of mountain you see on Earth.

✔ **Rays:** Bright lines formed by powdery debris thrown out from some impacts. They extend radially outward from young, bright impact craters, such as Tycho and Copernicus (see Figure 5-6).

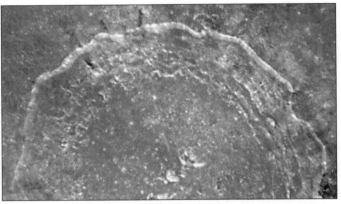

Figure 5-6:
A close-up view of the lunar crater Copernicus from the Hubble Space Telescope.

Courtesy of John Caldwell (York University, Ontario), Alex Storrs (STScI), and NASA

If you want to be able to distinguish one crater, rille, or lunar mountain range from another as you look through your telescope, you need a moon map or a set of lunar charts. These inexpensive items are available from astronomy and other scientific hobby supply houses and sometimes from map stores. Here are some good sources for these maps:

✔ Edmund Scientific (www.edsci.com) sells a full-color moon map poster (about $7) with feature identifications. The laminated version is better for use with your telescope; in the cool night air, unprotected paper can get wet with dew.

✔ Orion Telescopes & Binoculars (www.telescope.com) sells helpful guidebooks for lunar observing.

✔ Skyandtelescope.com offers the English-language edition of a highly regarded lunar guide, *Atlas of the Moon* by Antonin Rukl (about $45).

Remember, these maps and charts show only one side of the moon: the lunar near side.

For almost anything you want to see on the moon, the best viewing time is when the object is near the *terminator*, which is the dividing line between bright and dark. Details of lunar features are most evident when features are just to the bright side of the terminator. (A telescopic view is the closest you can get to the terminator unless you head for California to meet Governor Schwarzenegger or join the NASA astronaut corps.)

During a month, which is approximately the period of time from one full Moon to the next, the terminator moves systematically across the lunar near side so that at one time or another, everything you can see on the moon is close to the terminator. Depending on the time of the month, the terminator is either the place on the moon where the sun rises or the place where the sun sets. As you know from experience on Earth, shadows extend farther during sunrise or sunset and continually shrink as the sun gets higher in the sky. The length of the shadow when the sun is at a known altitude is related to the height of the lunar feature that casts it. The longer the shadow, the taller the feature.

About the worst time to look at nearly anything on the moon is during a full Moon. During a full Moon, the sun is high in the sky on most of the lunar near side, so the shadows are few and short. The presence of shadows cast by features on the moon helps you understand the *surface relief* — the way land-forms extend above or below their surroundings. But a full Moon isn't the time to seek relief.

Joining the dark side

You don't need a chart of the far side to help you observe the moon, because you can't see the far side; only the lunar near side is visible from Earth. Our view is limited because the moon is in *synchronous rotation,* meaning that it makes exactly one turn on its axis as it makes one orbit around Earth (the orbital period of the moon, which is the same as its "day," is about 27 days, 7 hours, and 43 minutes).

Astronomy supply houses and science stores sell moon globes, however, that depict the features of the entire moon, meaning the lunar near side and the far side. The lunar far side isn't a cartoon by Gary Larson; the Russians brought it to us. The Soviet space program first photographed the far side of the moon,

Going to lunar extremes: Bring your sunscreen, oxygen supply, and parka

When the sun is up, the temperature on the lunar surface goes up to as much as 243°F (117°C), but at night it drops to around −272°F (−169°C). These extreme temperature changes are due to the absence of any meaningful atmosphere to insulate the surface and reduce the amount of heat it loses at night. The moon has no liquid water. The surface is too hot, too cold, and too dry to sustain life as we know it. And there's no air to breathe.

which it did by snapping pictures with a robotic spacecraft very early during the Space Age. Since then, many different U.S. spacecraft, including the Lunar Orbiters and Clementine, have thoroughly mapped the moon.

Quite an impact: A theory about the moon's origin

Scientists know a lot about the ages of the rock in different terrains and parts of the moon. They acquired the data with radioactive dating of samples from the hundreds of pounds of lunar rocks that the six crews of NASA Apollo astronauts — who landed on the moon at different times from 1969 through 1972 — brought back to Earth.

Before the Apollo moon missions, several top experts confidently predicted that the moon would be the Rosetta stone of the solar system. With no liquid water to erode the surface, no atmosphere worth mentioning, and no active volcanism, they thought the surface should include plenty of primordial material from the birth of the moon and the planets. But the Apollo lunar samples threw rocks on their theory.

When a rock melts, cools, and crystallizes, all its radioactive clocks are reset. Radioactive isotopes begin producing fresh daughter isotopes that become trapped in the newly formed mineral crystals. The Apollo moon rocks show that the whole moon, or at least its crust down to a considerable depth, was melted well after 4.6 billion years ago. The very oldest surface rocks on the moon are *only* 4.5 billion years old. The difference between 4.6 and 4.5 billion years is 100 million years. And, unlike the minerals in earth rocks, which contain water bound up into the mineral structures, the moon rocks are bone dry.

The origin theory that has emerged to explain all this evidence, and to avoid the objections scientists posed against previous theories, is the *Giant Impact* theory. According to this theory, the moon is composed of material blasted out of the mantle of Earth by a huge object — with up to three times the mass of Mars — that struck young Earth a glancing blow. Some of the rock from the mantle of that long-vanished impacting object also was incorporated into the moon, according to the theory.

The giant impact on young Earth knocked all this material up into space as a vapor of hot rock. It condensed and solidified like snowflakes. The snowflakes knocked into each other and stuck together, and before you knew it, the moon had formed. It came together in powerful impacts of the last big pieces of accumulated rock, with the heat from each impact melting the rock.

All the impacts that caused the craters that we now see on the moon happened later, and most of them date back to more than 3 billion years ago.

The moon is less dense than Earth as a whole, and about as dense as Earth's mantle (the layer beneath the crust and above the core), according to this theory, because it was made from mantle material. (Density is a measure of the amount of mass that's packed into a given volume. If you have two cannonballs of the same size and shape, they have the same volume. But if one ball is made of lead and one is made of wood, the lead ball is heavier and has a higher density.) This theory predicts that the moon shouldn't have much of an iron core, if any. And a small core in a small object (meaning the moon) would have cooled and frozen long ago if it ever contained liquid iron. So the moon shouldn't be able to generate a global magnetic field. And that's exactly what space measurements tell us. The Lunar Prospector, a satellite put in orbit around the moon in the late 1990s, detected magnetic fields, but only at isolated places. The Lunar Prospector scientists concluded that the fields are fossil magnetic fields, produced in an unknown way, long ago.

The Giant Impact theory is currently our best guess. Unfortunately, we have no test for it at this time. For example, the theory predicts no special kind of rock that we could look for in the hundreds of pounds of lunar rocks that the Apollo astronauts collected. However, NASA, as of press time, is considering a future mission to the South Pole–Aitken Basin. In that huge crater, astronauts or robot rovers may find rocks knocked out from so deep inside the moon that they were beneath the surface layer that melted after the moon formed. Studies of those rocks may tell scientists if the Giant Impact theory is accurate.

And if scientists verify the Giant Impact theory, that will be one "giant step" for science.

Chapter 6

Earth's Near Neighbors: Mercury, Venus, and Mars

In This Chapter

▶ Meeting Mercury, the closest planet to the sun

▶ Checking out Venus, hot and stuffy with acid rain

▶ Discovering Mars, the planet we search for water and life

▶ Understanding what sets Earth apart

▶ Finding and observing our neighboring planets

You can spot Earth's neighboring terrestrial (or rocky) planets Mercury, Venus, and Mars with the naked eye and inspect them with your telescope. But they tantalize you by revealing only a little of their nature, which is why most of what scientists know about their physical properties, geologic forms, and likely histories is based on images and measurement data sent back to Earth by interplanetary spacecraft.

Mercury has played host to a single spacecraft, which flew past it three times and went on into space. Several probes have visited, orbited, and even landed on Venus. Mars has been the target of numerous probes, landers, and robot rovers, and NASA and other space agencies send more every two years. The mapping of Venus and Mars has been very thorough, but researchers still haven't seen large parts of Mercury.

In this chapter, I give you fascinating details about (and handy tips for viewing) Earth's closest neighbors in the solar system.

Hot, Shrunken, and Battered: Putting Mercury on a Platter

Despite three passes by the Mariner 10 spacecraft in 1973 and 1974, less than half of Mercury has been mapped. The remainder either wasn't in view from

Mariner 10 or was in darkness when it came by. To remedy this deficiency, NASA launched a space probe to Mercury on August 3, 2004. If all goes well, it will fly past and photograph the planet three times during 2008 and 2009 and then go into orbit around Mercury in 2011.You can follow the progress of the Mercury probe, called MESSENGER (MErcury Surface, Space ENvironment, GEochemistry, and Ranging), at the Web site `messenger.jhuapl.edu`.

To inspect the Mariner 10 images, you can visit the Mercury Mariner 10 Image Project at Northwestern University's Center for Planetary Sciences (`cps.earth.northwestern.edu/merc.html`). You can also check out the color section of this book to see an image of Mercury.

Here's what scientists know so far from info gathered mostly from Mariner 10 and from observations by radar astronomers on the ground who transmit pulses of radio waves toward Mercury and study the echoes:

- Mercury's surface is like that of Earth's moon (see Chapter 5), with one impact crater after another. (An impact crater is a hole in the ground caused by the fall of an asteroid, meteoroid, or comet.)
- Mercury has long, winding ridges that cut across impact craters and other geologic features. The ridges were probably caused by shrinkage of the crust, which was cooling from a molten state.
- Mercury has fewer small craters than the moon, in proportion to the number of large craters.

Highly cratered highlands are present on Mercury, as on Earth's moon (Mercury has no known moon of its own). But unlike the moon, Mercury's highlands are interrupted by gently rolling plains. Elsewhere, flat plains make up the Mercury lowlands.

The largest trace of any impact on Mercury is the Caloris basin. It isn't fully mapped because much of it was draped in darkness when Mariner 10 passed by. Astronomers' best estimates suggest that Caloris is about 830 miles (1,340 kilometers) across, which makes it among the largest impact basins in our solar system. Impact basins are huge craters, such as the lava-filled structures called *maria* on the moon. On the *antipode* of Caloris, which is the spot opposite Caloris on Mercury, is a strange region of broken hills and valleys. The collision that caused the Caloris basin generated powerful seismic waves, which traveled through Mercury and around its surface, converging at the antipode with a catastrophic effect.

Mercury has a density 5.4 times that of water. This high density means that Mercury has a huge iron core that constitutes the bulk of the planet. The outer layer of rock, called the *mantle,* must be no more than 380 miles (610 kilometers) thick. The presence of a global magnetic field, detected around Mercury by Mariner 10, suggests to many experts that some of that huge iron core must still be molten, although simple calculations indicate that the core should have cooled enough to solidify by now.

Faint traces of atmospheric gases exist on Mercury, but for practical purposes, the planet is airless like Earth's moon. It experiences extraordinary weather changes from day to night; temperatures can reach as much as 870°F (465.5°C) during the day and as low as –300°F (–184.4°C) at night. Areas of unusually high radar reflectivity near the North and South Poles may indicate a large amount of ice at the poles, in perpetually shadowed crater bottoms. MESSENGER will investigate whether this interpretation is correct.

Dry, Acidic, and Hilly: Steering Clear of Venus

Venus never sees a clear day; the planet is perpetually covered from equator to pole by a 9-mile-thick (15-kilometer) layer of clouds made up of concentrated sulfuric acid. And the surface has no relief from the heat: Venus is the hottest planet in the solar system, with a surface temperature of 870°F (465.5°C) that stays about the same from equator to pole, day and night.

And if the heat seems bad, check out the barometric pressure: It measures about 93 times the pressure at sea level on Earth. But forget about seas; you won't find any water on Venus. You can complain about the heat, but not the humidity — it's a dry heat, like in Arizona.

The bad news about the weather on Venus is that a perpetual rain of sulfuric acid falls all over the planet. The good news is that this rain is a *virga,* meaning rain that evaporates before it hits the ground.

Almost all the excellent images of the surface of Venus that you can find on NASA (and other) Web sites aren't photographs at all. What you see are detailed radar maps, notably from NASA's Magellan spacecraft. The clouds on the planet block the view of telescopes on Earth and of any camera on a Venus-orbiting satellite. The cloud tops are at an altitude of 40 miles (65 kilometers), much lower than where a satellite can operate.

The few images we have from Venus lander spacecraft, as pioneered by the former Soviet Union, show areas of flat rock plates, separated by small amounts of soil. The plates resemble areas of hardened basalt lava flows on Earth. But on Venus, the surface appears orange because the thick cloud cover filters the sunlight. You can see satellite radar maps and lander images of Venus and more at the Views of the Solar System Web site at www.solarviews.com/eng/homepage.htm. (You can see an image of Venus in the color section of this book, too.)

Flat plains that are volcanic lowlands with *rilles* (the winding canyons left by lava flows) cover the vast majority of Venus (about 85 percent). This territory includes the longest known rille in the solar system, Baltis Vallis, which stretches across Venus for about 4,230 miles (6,800 kilometers). Cratered highlands and deformed plateaus are also present.

Not as many craters dot Venus as you may expect, based on the number you see on Earth's moon (Venus has no known moon) and on Mercury. No small craters exist, and there aren't many large craters because Venus's surface was flooded with lava or reworked by volcanism (the eruption of molten rock from within a planet) after its bombardment by impacting objects had mostly ended. This flooding or reworking erased all or most of the early craters. Few large objects have struck Venus since the early craters were destroyed, and small objects don't make many craters on Venus, because objects capable of making craters up to 2 miles in diameter are impeded and destroyed by aerodynamic forces in the thick Venus atmosphere.

Huge volcanoes and mountain ranges cover the surface of Venus, but nothing resembling the nonvolcanic mountains on Earth (like the Rocky Mountains in the western United States or the Himalayas in Asia), which are caused by one crustal plate pushing into another. And Venus has no chains of volcanoes (like the Pacific "Ring of Fire"), which rise at the edges of plates. Plate tectonics and continental drift, as they occur on Earth, don't occur on Venus.

Red, Cold, and Barren: Uncovering the Mysteries of Mars

Scientists have topographically mapped Mars with high accuracy (topographically means that the altitudes of the landforms have been measured). You can find the National Geographic chart of the entire planet at the NASA Web site (ltp.gsfc.nasa.gov/tharsis/ngs.html). The map comes from an instrument called a laser altimeter on the Mars Global Surveyor (MGS), a satellite in orbit around Mars. A camera takes pictures from the MGS, and you can find its latest images at www.msss.com, the Web site of Malin Space Science Systems, a company that built and operated the camera.

While MGS was still monitoring Mars, another NASA spacecraft, the Mars Odyssey, arrived and began orbiting the planet in October 2001. You can see its findings at mars.jpl.nasa.gov/odyssey.

The European Space Agency doesn't get as much publicity as NASA, so you may not know that the Europeans have a Mars Express satellite that began orbiting the red planet on December 25, 2003. You can see splendid images from this spacecraft at www.esa.int/SPECIALS/Mars_Express.

Even though scientists have accurately mapped Mars, the planet still holds many mysteries that we want to solve. In the following sections, I cover theories about water and life on Mars. (And for even more on Mars, be sure to check out an image in the color section of this book.)

Where has all the water gone?

The topographic map of Mars shows that most of the Northern Hemisphere is much lower than the Southern Hemisphere. The huge northern lowland may be the bed of an ancient sea, but even if it isn't, strong evidence suggests that liquid water was once common on Mars.

Mars is cold and dry now, with a great deal of ice at the poles. By one estimate, enough ice is present to flood the entire planet to a depth of 100 feet if it melted. Some canyons on Mars look like a great flood carved them out, but not necessarily a planetwide flood. However, the polar ice won't melt; Mars is just too cold. The atmosphere is mostly carbon dioxide, and in winter, some of that gas freezes on the surface, leaving thin deposits of dry ice. At the pole where winter is underway, a thin cap of dry ice often tops the permanent cap of water ice. Dry riverbeds with streamlined islands and pebbles that look like they've been rounded in a torrent are among the other pieces of evidence for past liquid water on Mars. The pebbles were imaged with the Mars Pathfinder (which landed on Mars) and its little robot, Sojourner. Mars Odyssey, taking instrument readings from orbit, found likely evidence of large amounts of water, probably in a frozen state, just beneath the surface in large areas of Mars.

It does get comfortably warm on Mars at the equator, where noon-time temperatures can reach a balmy 62°F (16.6°C). However, don't stay the night — it can get down to as cold as –208°F (–133.3°C) after sunset. The seasons on Mars differ from Earth's seasons, too. On Earth (as I explain in Chapter 5), the seasons are caused by the tilt of Earth's axis with respect to the plane of Earth's orbit around the sun, *not* by changes in Earth's distance from the sun (which are negligible). On Mars, both the tilt of the planet's axis and the significant changes in its distance from the sun from one place in its orbit to another (because the orbit of Mars is more elliptical than Earth's almost-circular orbit) combine to produce "unearthly" seasons. Summer in the Southern Hemisphere on Mars is shorter and hotter than summer in the Northern Hemisphere, and winter in the Northern Hemisphere on Mars is shorter and warmer than winter in the Southern Hemisphere.

A magnetometer on MGS discovered long parallel stripes of oppositely directed magnetic fields frozen in the rocky crust of Mars. Mars doesn't have a global magnetic field today, but this finding may mean that it once had a global field that periodically reversed, just as Earth's field does (see Chapter 5), and it may also mean that Mars endured a crustal process resembling the seafloor spreading on Earth and producing a similar pattern. But the molten iron core on Mars must have frozen solid long ago, so a new magnetic field is no longer generated and the heat flow from the inside to the surface is so low that there's probably no volcanism still underway.

The volcanism that did occur on Mars produced immense volcanoes, such as Olympus Mons, which is about 370 miles (600 kilometers) wide and 15 miles (24 kilometers) high, or five times wider and almost three times higher than the largest volcano on Earth, Mauna Loa. Mars also has many canyons, including the immense Valles Marineris (Mariner Valley), which is 2,490 miles (4,000 kilometers) long. Impact craters dot the surface, too. The craters are more worn down than those on Earth's moon, because much more erosion has occurred, possibly caused by the water that produced the great floods on Mars (a controversial subject in astronomy to this day).

Does Mars support life?

People have many mistaken ideas about Mars, but some of the theories may actually be right; they just haven't been proven. These ideas all revolve around the possibility of life on Mars. Most of them are as improbable as the story about the future astronaut who returns from the planet: "Well, is there life on Mars?" the reporters demand. "Not much during the week," he says, "but on Saturday night . . ."

Claims about life strike out

The discovery of the "canals" on Mars spawned the first widespread speculation about the possibility of life. Some of the most famous astronomers of the late 19th and early 20th centuries were among those who reported the canals. Planetary photography wasn't very useful in those days, because the exposures were fairly long and atmospheric seeing (which I define in Chapter 3) blurred the images. So scientists believed that drawings by expert professional telescopic observers were the most accurate images of Mars. Some of these charts showed patterns of lines stretching and crisscrossing around the surface of Mars. Percival Lowell, an American astronomer, theorized that the straight lines were canals, engineered by an ancient civilization to conserve and transport water as Mars dried up. He concluded that the places where the lines crossed were oases.

Over the years, the idea of the "canals" and other reported indications of past or present life on Mars have struck out:

✔ When the American spacecraft Mariner 4 reached Mars in 1965, its photographs showed no canals, a conclusion verified in much greater detail by images from subsequent Mars probes. Strike one.

✔ Two later probes, the Viking Landers, conducted robotic chemical experiments on Mars to look for evidence of biologic processes such as photosynthesis or respiration. At first they appeared to have found evidence of biologic activity when water was added to a soil sample. But most scientists who reviewed the matter concluded that the water was reacting chemically with the soil in a natural process that doesn't involve the presence of life. Strike two.

✔ Viking Orbiters also sent back images of the Mars surface as they revolved around the planet. The images show, at one location, a crustal formation that — to some folks — looks like a face. Although many natural mountain peaks and stone formations on Earth resemble the profiles of the famous rulers, Native American tribal chiefs, and others for whom they're named, some true believers claim that the "face on Mars" is a monument of some type, erected by an advanced civilization. Later, sharper images from MGS showed that this landform doesn't look like a face at all. Strike three for the advocates of life on Mars.

But the idea of life was not "out," despite the three strikes.

The search for fossil evidence

In 1996, scientists analyzed samples of a meteorite that they believed came from Mars after being knocked off the planet by the impact of a small asteroid or comet. The scientists found chemical compounds and tiny mineral structures that they interpreted as chemical by-products and possible fossil remains of ancient microscopic life. Their work is very controversial, and many subsequent studies contradict these conclusions. Based on current research, scientists can't make a persuasive case that supports the theory of past life on Mars, nor can they disprove it.

The only thing to do is search systematically on Mars for evidence of life, past or present, in the regions that make the most sense — places where large quantities of water appear to have been present in the past and where layers of sediment were deposited in ancient lakes or seas. These types of places hold the most fossils on Earth.

Searches for sediments deposited by water in the past began on Mars in 2004 with NASA's Mars Exploration Rovers, called Spirit and Opportunity. They found plenty of persuasive evidence, including *blueberries,* or little round rocks that resemble known sedimentary features in the southwestern United States. You can see pictures and findings from the two rovers at `marsrovers.jpl.nasa.gov/home`.

Differentiating Earth through Comparative Planetology

Mercury is a tiny world of extreme temperatures, but it has a global magnetic field like Earth's, implying the presence of a molten iron core like Earth's. Although Venus and Mars don't have global magnetic fields, the planets are similar to Earth in many other ways. But liquid water and life occur today only on Earth as far as we know. What makes Earth different?

Venus, unlike Earth, has a hellish temperature. Venus is farther from the sun than Mercury but is even hotter. The high temperature is due to an extreme *greenhouse effect,* the process by which atmospheric gases raise the temperature by absorbing outward flowing heat. Earth's atmosphere may once have contained large amounts of carbon dioxide, the way Venus's atmosphere does now. But on Earth, the oceans absorbed much of the carbon dioxide, and that gas couldn't trap the heat the way it does on Venus.

Mars, on the other hand, is too cold to support life. Mars has lost most of its original atmosphere, and its current atmosphere isn't thick enough to produce a greenhouse effect sufficient to warm much of the surface above the freezing point of water very often or for very long.

The three large terrestrial planets are like the bowls of cereal in the child's story of Goldilocks. Venus and Mercury are too hot, Mars is too cold, but Earth is *just right* to support liquid water and life as we know it. Putting together the information on the basic properties of the terrestrial planets and their respective differences, scientists can conclude that

✔ Mercury is like the moon on the outside but like Earth on the inside.

✔ Venus is Earth's "evil twin."

✔ Mars is the little Earth that died.

Earth is the Goldilocks planet — just right!

When you contrast the properties of the planets like this, you can draw conclusions about their respective histories and why those different histories have brought the planets to their present conditions. Think that way and you're practicing what astronomers call *comparative planetology.*

Observing the Terrestrial Planets with Ease

You can spot Mercury, Venus, and Mars in the night sky with the aid of monthly viewing tips from astronomy magazines and their Web sites or with the aid of a desktop planetarium program (see Chapter 2). Venus is especially easy to find, because it's the brightest celestial object in the night sky next to our moon.

Mercury is the planet that orbits closest to the sun, and Venus is next. They both orbit inside the orbit of Earth, so Mercury and Venus are always in the same region of the sky as the sun as seen from Earth. Therefore, you can find these planets in the western sky after sunset or in the eastern sky before dawn. At those times, the sun isn't very far below the horizon, so you can see objects near and to the west of the sun in the morning before sunrise, and you can see objects near but east of the sun in the evening after sunset. Your motto as a Mercury or Venus spotter should be to "look east, young woman" or "look west, young man," depending on whether you skywatch at dawn or dusk and whether you're a fan of old western movies.

A bright planet appearing in the east before dawn is commonly called a *Morning Star,* and a bright planet in the west after sunset is an *Evening Star.* As Mercury and Venus move swiftly around the sun, this week's Morning Star may be the same object as next month's Evening Star (see Figure 6-1).

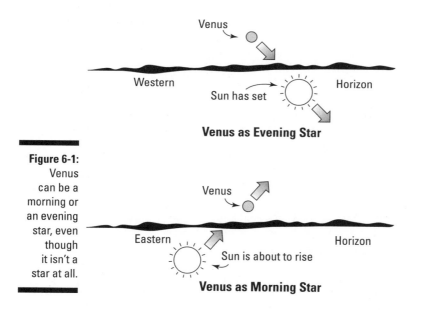

Figure 6-1:
Venus
can be a
morning or
an evening
star, even
though
it isn't a
star at all.

In the following sections, I explain the best time to observe the terrestrial planets based on elongation, opposition, and conjunction — three terms that describe the planets' positions in relation to the sun and Earth — and how to use this understanding in your observations of them. (I list the planets in the order of ease of observation, starting with Venus, the easiest.)

Understanding elongation, opposition, and conjunction

Elongation, opposition, and conjunction are terms that describe a planet's position in relation to the sun and the earth. You encounter these terms when you check listings of the planets' positions in order to plan your observations. Here's what the terms mean:

- *Elongation* is the angular separation between a planet and the sun, as visible from Earth. Mercury's orbit is so small that the planet never gets more than 28 degrees from the sun. During some periods, it doesn't get farther than 18 degrees from the sun, making it hard to spot. Venus can get up to 47 degrees from the sun.

  Greatest western (or eastern) elongation occurs when a planet is as far from the sun as it can get during a given *apparition* (when the planet is visible from Earth). Some greatest elongations are greater than others, because sometimes Earth is closer to the planet than at other occasions. Elongation is especially important when you observe Mercury, because the planet is usually so close to the sun that the sky at its position isn't very dark.

- *Opposition* occurs when a planet is on the opposite side of Earth from the sun. This never happens for Mercury or Venus, but Mars is at opposition about once every 26 months. This is the best time to observe the planet, because it appears largest in a telescope. And, at opposition, Mars is highest in the sky at midnight, so you can view it all night.

- *Conjunction* occurs when two solar system objects are near each other in the sky, such as when the moon passes near Venus as we see them. In reality, Venus is far beyond the moon, but we see a conjunction of the moon and Venus.

Conjunction has a technical meaning as well. Instead of describing positions in *right ascension* (the position of a star measured in the east-west direction) and *declination* (the position of a star measured in the north-south direction), astronomers sometimes use ecliptic latitude and longitude. The ecliptic is a circle in the sky that represents the path of the sun through the constellations. *Ecliptic latitude* and *longitude* measure degrees north and south (latitude) or east and west (longitude) with respect to the ecliptic. (Don't worry; you won't need to use the ecliptic system when you observe the terrestrial planets. But knowing about it helps you understand the definitions of superior and inferior conjunction that follow.)

You need to master some tricky terminology in order to understand conjunctions and oppositions; namely, the labeling of planets as superior and inferior and the labeling of conjunctions as superior and inferior. A *superior planet* orbits outside the orbit of the Earth (so Mars is a superior planet, for example). An *inferior planet* orbits inside the orbit of the Earth (so Mercury and Venus are inferior planets; in fact, they're the only inferior planets).

When a superior planet is at the same longitude as the sun as seen from Earth, it's on the far side of the sun and is said to be at *conjunction* (see Figure 6-2). When that same planet is on the opposite side of Earth from the sun (also shown in Figure 6-2), it's at *opposition*.

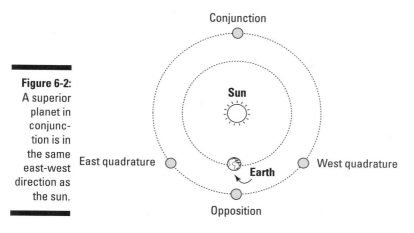

Figure 6-2:
A superior planet in conjunction is in the same east-west direction as the sun.

Conjunction and opposition diagram showing East quadrature, West quadrature, Sun, and Earth.

Conjunction is a bad time to observe a superior planet, because its position is on the far side of the sun. So don't try to observe Mars at conjunction; you won't see it. The best time to observe Mars is at opposition.

A superior planet has conjunctions and oppositions, but an inferior planet has two kinds of conjunction and never has an opposition (see the diagram in Figure 6-3). When the inferior planet is at the same longitude as the sun and is between the sun and Earth, it's at inferior conjunction. And when the inferior planet is at the same longitude as the sun but is beyond the sun as seen from Earth, the planet is at superior conjunction.

If you can explain all that to your friends, you'll truly feel superior! Feel free to do the explaining "in conjunction" with Figures 6-2 and 6-3.

You can see Venus best at inferior conjunction, when it looks largest and brightest, but Mercury is too close to the sun for viewing at inferior or superior conjunction; the best viewing time is greatest elongation.

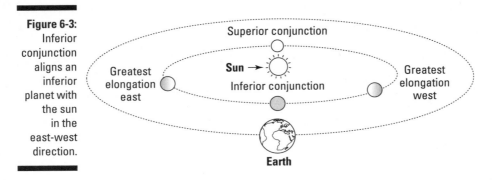

Viewing Venus and its phases

The simplest planet to find is Venus. The second rock from the sun is so bright that people with no knowledge of astronomy frequently notice it and call radio stations, newspapers, and planetariums to ask what "that bright star" is.

When scattered clouds move from west to east in front of Venus, inexperienced viewers often misunderstand the scene. Folks think that Venus (which they don't recognize) is moving rapidly in the opposite direction of the clouds. Due to its brightness and the mistaken impression that it's moving rapidly behind a cloud deck, people often report Venus as an unidentified flying object. It isn't. Astronomers know it well.

After you become familiar with Venus, you may be able to spot it in broad daylight. Often, Venus is bright enough that if the sky is clear, you can spot it in the daytime by using *averted vision*. That means that you can glimpse it "out of the corner of your eye." For some reason, you may be able to spot a celestial object easier with averted vision than if you look right at it. This may be a survival trait; it makes it a little harder for an enemy or a predator to sneak up on you from the side.

A small telescope can show you Venus's most recognizable features: its phases and changes in apparent size. Venus has phases similar to Earth's moon (see Chapter 5), and for the same reason: Sometimes part of the Venus hemisphere that faces the sun (and is therefore bright) is directed away from Earth, so a telescopic view of Venus shows a partly illuminated and partly dark disk.

The dividing line between the bright and dark parts of Venus is called the *terminator,* just as on the moon. Don't worry: This terminator is just a perfectly safe imaginary line on Venus (see Chapter 5).

Wait just an arc minute (or second)

Scientists measure apparent sizes in the sky in angular units. Something that goes all the way around the sky, such as the Celestial Equator, is 360 degrees long. The sun and the moon, in comparison, are each about a half-degree across. The planets are much smaller, so smaller units are necessary to describe them. A degree is divided into 60 minutes of arc, and a minute of arc — called an *arc minute* or *arc min* — is divided into 60 seconds of arc — usually abbreviated as *arc second* or *arc sec*. A degree is made up of 3,600 (60 times 60) seconds of arc. In many astronomy books and articles, a single prime symbol (') represents a minute of arc, and a double prime (") represents a second of arc. Readers often mistake these symbols as abbreviations for feet and inches. You can tell when a clueless copy editor has had the last cut at an astronomy article, because you see a statement like "The moon is about 30 feet in diameter."

Venus is actually only about 5 percent smaller in diameter than Earth. Its apparent size of angular diameter ranges from about 10 arc sec, when Venus is farthest away (and with a full Moon shape), to about 58 arc sec in diameter when it's closest (and is a narrow crescent).

As Venus and Earth orbit the sun, the distance between the two planets shifts substantially. At its closest to Earth, Venus is a mere 25 million miles away, and at its farthest, it's a full 160 million miles distant. What's important here is the proportional change: At closest approach, Venus is about six times nearer to Earth than at its most distant position. And it looks six times bigger through a telescope.

What you don't see when you view Venus are striking features, like the Man in the Moon. Venus is totally covered by thick clouds, and all you can see is the top of the clouds. Venus is so bright because it orbits relatively close to both the sun and Earth and because it has a bright, reflecting cloud layer. But sometimes you may be able to discern the horns of the Venus crescent extending farther into the dark side than predicted for the phase on that day. What you see is some sunlight that has bounced around in Venus's atmosphere and passed beyond the terminator into the side of the planet where night has fallen.

Images of Venus with striking cloud patterns, like those you see in books, were made in ultraviolet light where the patterns show up. Ultraviolet light doesn't pass through our atmosphere (hurray for the ozone layer, which blocks this hazardous radiation), so you can't view Venus in it. In fact, you can't see ultraviolet light anyway; it's invisible to the human eye. But telescopes on satellites and space probes above or beyond the atmosphere can take ultraviolet pictures.

> # Waiting to catch the Venus transit
>
> One of the rarest planetary events you can see is a *transit* of Venus, when the planet passes right in front of the sun and appears as a tiny black disk against the bright solar surface.
>
> You can see this event with the naked eye (be sure to use a safe solar filter, as I describe for sunspot observing in Chapter 10), but you have only one chance (unless you already saw the transit of Venus on June 8, 2004), and that's on June 6, 2012. After that, you can't see a transit of Venus until the year 2117.
>
> The transit on June 6, 2012 will be visible, for its full duration or in part, from much of Earth but not at all in Portugal, southern Spain, west Africa, and South America's southeastern two-thirds, according to NASA transit expert Fred Espenak (`sunearth.gsfc.nasa.gov/eclipse/transit/venus0412.html`).

On rare occasions, observers report a pale glow on the dark part of Venus. This glow, called the *ashen light,* is sometimes a real phenomenon and sometimes a trick of the imagination. After centuries of study, experts still can't explain the ashen light, so some of them even deny that it exists. But with luck, you may see it. People claim to see other features on Venus through their telescopes, but almost all the reports are wrong. Experiments show that the reports are usually due to a psychological effect: If people view a featureless white globe from a distance, they may discern patterns that don't exist.

Watching Mars as it loops around

Mars is a bright red object, but it isn't nearly as dazzling as Venus. So check your sky maps to make sure that you don't mistake a bright red star, such as Antares in Scorpius (whose name means "rival of Mars"), for the red planet.

The great advantage in observing Mars is that when it appears in the night sky, it often remains visible for much of the night, unlike Mercury and Venus, which set fairly soon after sunset or rise only shortly before dawn. You usually have time for dinner and the nightly news before you head into the backyard to check on Mars.

With a small telescope, you can spot at least a few dark markings on Mars. The best periods to see the features last a few months but occur only about every 26 months, when Mars is at opposition. At opposition, Mars looks biggest and brightest, and you can see details more easily on its surface.

The upcoming oppositions of Mars are in

November 2005

December 2007

January 2010

March 2012

April 2014

Don't miss them!

At its best oppositions — when it looks biggest and brightest — Mars is south of the Celestial Equator, but you can still observe it from temperate latitudes in the Northern Hemisphere.

The easiest Martian surface feature to spot with a small telescope is usually Syrtis Major, a large dark area extending northward from the equator. Mars's day is nearly the same as Earth's: 24 hours, 37 minutes. So if you look at Mars off and on during a night, you may be able to see Syrtis Major move slowly across the planet's disk as Mars turns. Experienced amateur planetary observers may see its polar caps and other markings as well.

Tracking Mars's backtracking

A basic project for beginning planet gazers is to track the motion of Mars across the constellations; all you need are your eyes and a sky map.

Locate Mars among the stars and mark that position with soft pencil on your map. If you repeat this observation on each clear night, you can see a pattern emerge that puzzled the ancient Greeks and led to complicated theories — most of them wrong.

Most of the time, Mars moves eastward from night to night, just as Earth's moon moves eastward across the constellations. The moon keeps going, but Mars sometimes reverses course. For two to almost three months (62 to 81 days) at a time, Mars heads west across the constellations, moving backward for 10 to 20 degrees. After this period, it gets back on track and heads east again. The backtracking is called Mars's *retrograde motion*.

The backtracking isn't a case of Mars not knowing whether to come or go. The retrograde motion is just an effect produced by Earth racing around the sun. While you chart Mars's motion, you stand on Earth, which races around the sun once every 365 days. Mars moves slower, making one full orbit in 687 days. As a result, when we pass Mars on our inside track (lapping it), Mars seems to move backward against the reference frame of the distant stars. But in reality, Mars always forges ahead.

NASA images of Mars, taken by interplanetary probes and the Hubble Space Telescope, are much too detailed to guide you in small-telescope observation. You need a simple *albedo map,* which charts and names the bright and dark areas on Mars as visible with small telescopes. An albedo map offers more detail than the average observer ever sees, and it offers a good guide and a challenge to your observing skills. You can find such a map in *Norton's Star Atlas and Reference Handbook* (see Chapter 3) or on the Web site of the Mars Section of the Association of Lunar and Planetary Observers, at www.lpl. arizona.edu/~rhill/alpo/mars.html. I also recommend *A Traveler's Guide to Mars,* a quality paperback with foldout charts, by William K. Hartmann, a leading planetary scientist and a space artist to boot (Workman Publishing).

Astronomers rate sky conditions in terms of *seeing* (the steadiness of the atmosphere above the telescope), *transparency* (the freedom from clouds and haze), and *sky darkness* (the freedom from interfering artificial light, moonlight, or sunlight). When observing a bright planet such as Mars, good seeing is the most important factor, and dark sky is least important. But the darker the sky, the steadier the air, and the higher the transparency, the more you can enjoy the night.

With good seeing, the stars don't twinkle quite as much, and you can use a higher magnification eyepiece with the telescope to bring out fine details on Mars or another planet. When the seeing isn't very good, the telescopic image is blurred and seems to jump around. Under adverse conditions, high magnification is useless; you only magnify the blurred, jumping image. Use a low-power eyepiece for the best results.

Unfortunately, even when atmospheric conditions are ideal at your observing site and when an opposition of Mars is in progress, disaster may strike. Mars is a planet that experiences worldwide dust storms, which hide its surface features from view.

In fact, professional astronomers sometimes rely on amateur astronomers to help monitor Mars, to let them know when a dust storm begins, and to report other pronounced changes in the appearance of the planet. You can get information on this program at the International MarsWatch 2003 Web site (elvis. rowan.edu/marswatch). You have more fun with a sharp view of Mars, but if all else fails, at least you may get credit for discovering a dust storm. The experts welcome your dust report instead of brushing it off.

You need experience to become a reliable telescopic Mars observer. As a beginning observer, don't assume that a great dust storm is in progress just because you can't make out any detail. Get accustomed to seeing Mars in detail. Only then should you consider that, when you can't see details, the planet is at fault, not your inexperience. Remember this scientific motto: "The absence of evidence is not necessarily evidence of absence." You may not see detail the first time you look, but that doesn't mean that a dust storm is obscuring your view. As a telescopic observer, you have to train your viewing skills, just as gourmets and wine lovers train their palates.

And just for your information: Mars has only two known moons, Phobos and Deimos. These tiny celestial bodies aren't visible with small telescopes.

Outdoing Copernicus by observing Mercury

Historians say that the great 17th-century Polish astronomer Nicholas Copernicus, who proposed the *heliocentric* (sun-centered) *theory* of the solar system, never spotted the planet Mercury.

But Copernicus didn't have modern aids, such as desktop planetariums, astronomy Web sites, and monthly astronomy magazines (see Chapter 2). You can use these aids to find out when Mercury will be best placed for observation during the year: the times of greatest western and eastern elongation (terms that I cover in the section "Understanding elongation, opposition, and conjunction" earlier in this chapter), which occur about six times each year.

At temperate latitudes, such as those of the continental United States, Mercury is usually only visible in twilight. By the time the sky is dark, well after sunset, Mercury has set, too. And in the morning, you can't spot Mercury until the impending dawn starts to light the sky. Mercury resembles a bright star but appears much dimmer than Venus in the west at dusk or in the east before dawn.

You can get more information on observing Mercury and the other planets from the Association of Lunar and Planetary Observers (ALPO), which also collects sketches and other planetary observations made by amateur observers and offers observing forms, charts, and other publications. Some of the association's advice is more optimistic than mine concerning what's visible in small telescopes, but why not try (to paraphrase a slogan of the U.S. Army) to "see all that you can see." The ALPO home page is at `www.lpl.arizona.edu/alpo`.

Hopping aboard the Mercury transit

Like Venus, you sometimes see Mercury in transit, when it passes across the face of the sun and appears as a small black disk against the solar surface as visible from Earth. Observe a transit of Mercury with a telescope, using the procedures for safe solar viewing that I explain in Chapter 10. (Remember, you're viewing Mercury against the sun, so solar viewing precautions are absolutely necessary.) Two transits of Mercury will occur in the next decade: on November 8, 2006, and on May 9, 2016. Depending on where you live, you may have to travel to see one of them, although you can usually see a transit from a large portion of the earth.

Be an early riser for Mercury

Mercury is much smaller than Venus, but you can see its phases through your telescope. The best time to do this is when Mercury is at western elongation and appears in morning twilight. The atmospheric steadiness or seeing is almost always better low in the east near dawn than it is low in the west after sunset, so you get a much sharper view in the morning. Standard guides, like the highly regarded *Observer's Handbook* (an annual publication of The Royal Astronomical Society of Canada; www.rasc.ca), the *Astronomical Calendar* (published annually by Universal Workshop; www.universalworkshop.com), and the astronomy magazines and their Web sites (see Chapter 2), all tell when the elongations are due.

You need a viewing site with a clear eastern horizon, because Mercury doesn't get very high in the sky when the sun is below the horizon. If you have trouble spotting it with your naked eye, sweep around that part of the sky with a pair of low-power binoculars. And if you have a computerized telescope with a built-in database, you can just punch in "Mercury" and let the telescope do the finding.

Don't expect to see surface markings

Seeing surface markings on Mercury with a small telescope, or with almost any telescope on Earth, is extremely difficult. Mercury's apparent size at greatest elongation is only about 6 to 8 arc sec (see the sidebar "Wait just an arc minute (or second)," earlier in this chapter, for details).

Some experienced amateur observers report seeing surface markings, but no useful information has ever come from such sightings. A few of the greatest planetary observers of all time thought that they could see and draw the surface markings. From their drawings, the observers tried to deduce the rotation period or "day" of Mercury. The experts believed that Mercury's day was equal to the length of the year on Mercury, or 88 Earth days. But they were wrong. Radar measurements later proved that Mercury turns once every 59 Earth days.

Nevertheless, after you find out how to spot Mercury by eye and then check on its phases with your telescope, you'll be way ahead of Copernicus!

Mercury lovers choose morning

Here's why the seeing is better near the dawn horizon than near the sunset horizon: By sunset, the sun has warmed Earth's surface all day, so as you look out low in the sky to the west, you look through turbulent currents of warm air rising from the surface. But in the morning, Earth has had all night to cool down and stabilize. It takes a few hours for the rising sun to warm the land and mess up the seeing again.

Chapter 7

Rock On: The Asteroid Belt and Near-Earth Objects

In This Chapter

▶ Discovering basic facts about asteroids

▶ Evaluating Earth's risk of a dangerous asteroidal impact

▶ Observing asteroids in the night sky

*A*steroids are big rocks that circle the sun. The vast majority of asteroids are safely beyond the orbit of Mars in an area called the *Asteroid Belt,* but thousands of asteroids follow orbits that come close to or cross Earth's orbit. Many scientists believe that an asteroid hit Earth about 65 million years ago, wiping out the dinosaurs and many other species.

In this chapter, I introduce you to these big rocks and explain the best ways to observe them. And, in case you're worried, I tell you the truth about the risk of an asteroid hitting Earth in the future and fill you in on the research scientists are conducting to deal with the possibility.

Taking a Brief Tour of the Asteroid Belt

Asteroids are also called minor planets, because when they were first discovered, experts thought that they were objects like the planets. But astronomers now believe that asteroids are remnants of the formation of the solar system — objects that never merged with additional space debris to grow into planets. Some asteroids, such as Ida, even have their own moons (see Figure 7-1). Asteroids are made of silicate rock, like the rocks of Earth, and of metal (mostly iron and nickel). Some asteroids may also contain carbonaceous (or carbon-bearing) rock.

The only definition for an asteroid is that it's a small body in the solar system, made of rock and iron — a definition that also includes meteoroids. So the biggest meteoroids and the smallest asteroids are indistinguishable from one another.

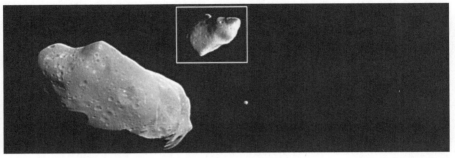

Figure 7-1:
The asteroid Ida has its own moon, Dactyl.

Courtesy of NASA

Most of the known asteroids are in a huge, flat region centered on the sun and located between the orbits of Mars and Jupiter. We call this region the *Asteroid Belt*. Asteroids range in size from huge asteroids such as Ceres, which is 580 miles (933 kilometers) in diameter, to large meteoroids, which are just fragments of asteroids (see Chapter 4). A boulder-sized space rock is a very small asteroid or a very large meteoroid; feel free to take your pick.

Table 7-1 lists the four biggest objects in the Asteroid Belt. The two largest, Ceres and Pallas, are at nearly the same average distance from the sun, although Pallas has a much more elliptical orbit.

Table 7-1	The "Big Four" of the Asteroid Belt		
Name	*Diameter in Miles*	*Diameter in Kilometers*	*Mean Distance from the Sun (A.U.)*
Ceres	580	934	2.77
Pallas	327	526	2.77
Vesta	317	510	2.36
Hygiea	254	408	3.14

As of February 1, 2005, there were about 264,000 known asteroids, of which 12,136 have been named (including one that the International Astronomical Union was kind enough to name after me). Most were found in recent years by robotic telescopes designed for the purpose, but experienced amateur astronomers who mount advanced digital cameras on their telescopes are also making discoveries.

You can readily see the largest asteroids, such as Ceres and Vesta, through small telescopes (see the later section "Searching for Small Points of Light" for more about observing asteroids).

Ceres and Vesta are so big that their own gravity makes them round. But smaller asteroids are often potato-shaped and frequently look blasted apart (see Figure 7-2) because, indeed, they have been. The asteroids in the belt constantly bump into each other, making impact craters and breaking off big and little chips. The big chips are simply smaller asteroids, and the little chips are asteroidal meteoroids.

At rare intervals, small asteroids (or large meteoroids) have smashed into Earth (see the next section for more about this phenomenon). Asteroid impacts (and comet impacts) have also covered the moon with craters (as I describe in Chapter 5).

Asteroids have craters too, but they're much harder to see with telescopes because asteroids are so small. In most telescopes, an asteroid is just a point of light, like a star. You can glimpse a large crater on Vesta in an animation of Hubble telescope photos at hubblesite.org/newscenter/newsdesk/archive/releases/1997/27/video/a.

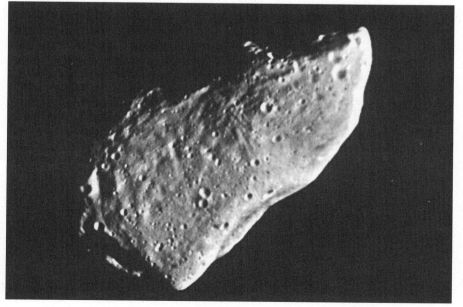

Figure 7-2:
Some asteroids resemble big potatoes.

Courtesy of NASA

Understanding the Threat That Near-Earth Objects Pose

Not all asteroids are orbiting safely beyond Mars. Thousands of small asteroids follow orbits that cross or come close to Earth's orbit. Astronomers call these neighbors *Near Earth Objects (NEOs),* and they've classified 624 of them as *Potentially Hazardous Asteroids (PHAs),* as of February 2005. Some day one of these scary neighbors may come uncomfortably close to Earth or even strike our planet. The Minor Planet Center (MPC) of the International Astronomical Union keeps tabs on PHAs, and several observatories sweep the skies to discover more of them.

The MPC's Web site (cfa-www.harvard.edu/iau/mpc.html) offers a mix of information for experts and amateur astronomers alike, including maps of the inner and outer solar system, updated daily, that show where the planets and many of the asteroids are located in space.

Astronomers don't know of any specific object that's currently a menace to Earth. Conspiracy theorists think that if astronomers did know about a doomsday asteroid, we wouldn't tell. But face it, if I knew the world was in danger, I'd settle my affairs and head for the South Seas instead of sitting around finishing this chapter!

In 1998, the Hollywood movies *Armageddon* and *Deep Impact* gave sensationalized versions of what could happen if a large asteroid or comet was on a collision course for Earth. Such catastrophe stories are inspired, in part, by the widely accepted conclusion that an asteroid about 6 miles (10 kilometers) wide struck Earth about 65 million years ago. The Chicxulub crater, a 110-mile-wide (180-kilometer-wide) geologic formation resting partly on Mexico's Yucatan peninsula and partly offshore in the Gulf of Mexico, may be the surviving trace of the impact, which is theorized to have wiped out the dinosaurs. (It certainly didn't do them any good; if you want to see a "live" dinosaur nowadays, renting a video of *Jurassic Park* is the closest you can come.)

The action of weather and geological processes, such as mountain building, erosion, and volcanism, has eroded the impact craters on Earth and eliminated most of them. You can see aerial photos of over 170 of our planet's remaining craters at the Earth Impact Database site of the University of New Brunswick, at www.unb.ca/passc/ImpactDatabase/images.html.

One impact by a small asteroid caused the famous Meteor Crater (which ought to be called Meteoroid Crater or Asteroid Crater) in northern Arizona, near Flagstaff. The site is well worth a visit because it's the largest and best-preserved impact crater on Earth.

For a brief time in March of 1998, many people feared that a small, newly discovered NEO could strike Earth in the year 2028. Astronomers eliminated that possibility within a day when additional observations showed that the asteroid's orbit won't intersect Earth. And some experts even disagreed with the initial prediction — as experts often do.

Although Earth appears to be safe for now, scientists may discover an NEO on a collision course with Earth in the future, so they spend time now studying the options for what can be done in such an eventuality.

When push comes to shove: Nudging an asteroid

Some experts propose the development of a powerful nuclear missile to intercept a killer asteroid before it can strike the earth. But if we blow up an asteroid heading our way, the results could be worse than the damage caused by the impact of the intact asteroid. It would be like the scene in the Walt Disney movie *Fantasia,* in which the sorcerer's apprentice chops up the out-of-control magic broom that won't stop fetching water. He creates a whole bunch of little brooms that each start fetching water.

If we blow up an asteroid with a nuclear bomb, a swarm of smaller rocks — rather than one big rock heading for Earth — would follow the same deadly trajectory. The rocks would pack more wallop than all the Pentagon's weapons combined. A better idea is to use the nuclear missile (or perhaps some other kind of missile) to only nudge the asteroid so that it passes through its intersection point with Earth's orbit a little early or a little late, when Earth isn't at that spot or has already gone past. Phew!

The problem with nudging an asteroid is that scientists don't know how much force to apply. We don't want to break it up, but because the mechanical strength of asteroids is unknown to us, we don't know how hard to hit it. Asteroids may be made of strong rock or fragile stone. Some may be mostly solid metal. If we don't know our enemy, we can make things worse by striking in the wrong way.

In *Fantasia,* the sorcerer himself breaks the spell on the enchanted broom, but without a sorcerer to make the asteroids disappear, we need hard information to design a system that can safely protect Earth from asteroids.

Forewarned is forearmed: Surveying NEOs to protect Earth

Astronomers have a plan to help design a system that can protect Earth from renegade asteroids:

1. **Take a census of the NEOs to make sure that we've located every rock that measures a kilometer or more in size in our region of the solar system.**

   NEOs of this size can become PHAs if their orbits take them close to Earth.

2. **Track these NEOs and compute their orbits to determine if any are likely to strike Earth in the foreseeable future.**

3. **Study the physical properties of asteroids to discover as much as we can about them.**

   For example, make telescopic observations to determine what kind of rock or metal they are made of.

4. **When astronomers understand the threat, an engineering team can design a space mission to counteract it.**

To survey the NEOs, special-purpose asteroid discovery telescopes are in operation at several locations. You can visit their Web sites and see their recent findings. Two of the most important projects are

- ✔ The Lincoln Near Earth Asteroid Research (LINEAR) project at White Sands, New Mexico, funded by the U.S. Air Force; www.ll.mit.edu/ LINEAR

- ✔ NASA's Near-Earth Asteroid Tracking (NEAT) project, observing from the Maui Space Surveillance Site in Hawaii and Palomar Observatory in California; neat.jpl.nasa.gov

A private organization, the Spaceguard Foundation, is dedicated to saving Earth from killer asteroids. It may have bitten off more than it can chew; just saving the whales or the spotted owls is hard enough. But you can look the organization up and join at the Web site cfa-www.harvard.edu/~marsden/SGF.

A list of PHAs is maintained by the Minor Planet Center at the Web site cfa-www.harvard.edu/iau/lists/Dangerous.html. Probably few if any of these asteroids are larger than about 10 miles in diameter. But a rock a few miles in size that strikes Earth at 25,000 miles per hour (11 kilometers per second) would be a far greater catastrophe than the simultaneous explosion of all the nuclear weapons ever made. That would be a rare case when astronomy isn't fun.

Searching for Small Points of Light

Looking for asteroids is like scanning the sky for comets (see Chapter 4), except that you look for a small point of light resembling a star rather than a fuzzy image. But unlike a star, an asteroid moves perceptibly against the background of other stars from hour to hour and from night to night.

You can easily see the largest asteroids, such as Ceres and Vesta, through small telescopes; astronomy magazines publish short articles and sky charts to guide you in advance of good viewing periods (in general, there are no best times of the day or year to view asteroids). Most good planetarium programs also make sky maps that point out the location of asteroids. (See Chapter 2 for more about magazines and planetarium programs and Chapter 3 to find out about telescopes.)

You aren't ready to search systematically for unknown or "new" asteroids until you become a skilled amateur astronomer with a few years of experience. Advanced amateurs search for new asteroids with electronic cameras on their telescopes. They collect a series of images of selected areas of the sky, generally in the direction opposite the sun (which, of course, is below the horizon). When they see a small point of light (resembling a star) change its location, they probably see an asteroid.

The easiest asteroid-related activity for beginners to try is observing occultations. An *occultation* is a kind of eclipse that occurs when a moving body in the solar system passes in front of a star. The bodies responsible can be Earth's moon (lunar occultation), the moons of other planets (planetary satellite occultation), asteroids (asteroidal occultation), or planets (planetary occultation). The rings of planets and comets can also cause occultations. An occultation doesn't look like much; you just see the star disappear for a short time during the eclipse.

The following sections tell you how to time and track asteroidal occultations.

Timing an asteroidal occultation

You can enjoy an asteroidal occultation without obtaining scientific data, but what a waste of a unique opportunity! The details of an occultation differ from place to place on Earth. For example, the same occultation may last longer as seen from one place on Earth than from another, or it may not occur at a certain location. So at some locations, you see the star being eclipsed, and at other locations, you see the star without an eclipse. From occultation data, astronomers can get a more accurate picture of a number of sky objects. For example, sometimes the occultations reveal that what seems to be an ordinary star is actually a close *binary system* (two stars in orbit around a common center of mass; see Chapter 11 for details on binary stars).

To make your asteroidal observation scientifically useful, you need to time it accurately and know the exact location (latitude, longitude, and altitude) of the place where you are when you observe the occultation. In the past, observers figured out their location by consulting topographic maps. But nowadays, if you observe in a group, someone probably has a GPS (Global Positioning System) receiver, such as those used by boat operators and private pilots. You can purchase them for as little as $100.

You should report your observations to the International Occultation Timing Association (IOTA). Check out its Web site at www.lunar-occultations.com/iota/iotandx.htm. It includes an "Asteroidal Occultation Report Form" that you can fill out and submit online. The IOTA Web site is regularly updated to provide the latest predictions of occultations by asteroids and other objects.

Helping to track an occultation

Asteroidal occultations are much trickier to observe than lunar occultations, because astronomers often can't predict them with sufficient precision. Astronomers go to various spots on the predicted *occultation ground track* (a narrow band across the surface of the earth where astronomers expect the occultation to be visible — just like the path of totality in an eclipse of the sun, which I describe in Chapters 2 and 5) and attempt to observe asteroidal occultations. But because the diameters, orbits, and shapes of most asteroids aren't known with sufficient accuracy, the predictions can't be precise. Because the occultation may be visible at some places and not at others, astronomers need volunteers to monitor an asteroidal occultation at many locations. Amateur observations help determine the sizes and shapes of the asteroids involved in the occultations. You can join too. Contact IOTA through the Web site www.lunar-occultations.com/iota/iotandx.htm.

IOTA recommends that you begin occultation study by observing with an experienced astronomer, just to get the hang of it. I bet you'll enjoy every iota of it!

Chapter 8

Great Balls of Gas: Jupiter and Saturn

In This Chapter

▶ Understanding the recipe for gas-giant planets

▶ Spotting the Great Red Spot and Jupiter's moons

▶ Observing Saturn's rings and moons

*J*upiter and Saturn, located beyond Mars and the Asteroid Belt, are among the best sights to see through a small telescope, and at least one is usually well placed for observations. The four largest moons of Jupiter and the famous rings of Saturn are the favorite targets when amateur astronomers give friends and family some peeks through their telescopes. And although you may not be able to tell through the telescope, the underlying science of these huge planets and their satellites is fascinating, too. In this chapter, I describe the magnificent sights you can observe through your telescope and clue you in to the basic facts about the two largest planets in our solar system.

The Pressure's On: Journeying Inside Jupiter and Saturn

Jupiter and Saturn are like hot dogs with unapproved food coloring. The meat isn't the mystery, the additives are. What you see in telescopic photographs of Jupiter and Saturn are the clouds, made of ammonia ice, water ice (like the cirrus clouds on Earth), and a compound called ammonium hydrosulfide. Water-drop clouds may be a part of the mix, too. But the appearances are deceiving. These cloud materials are made of trace substances. Jupiter and Saturn are mostly hydrogen and helium, like the sun. And, despite much theorizing, scientists have no proof of what makes the Great Red Spot on Jupiter red or of what produces the other off-white tints in the clouds of the two great planets.

Jupiter and Saturn are the largest of the four gas-giant planets (the others are Uranus and Neptune). Jupiter has 318 times the mass of Earth; Saturn surpasses Earth's mass by about 95 times. As a result, their gravity is enormous, and inside the planets, the weight of the overlying layers produces enormous pressure. Descending into Jupiter or Saturn is like sinking in the deep sea. The farther down you go, the higher the pressure. And unlike the sea, the temperature increases radically with the depth. Don't even think about scuba diving there.

Up at the atmospheric levels where astronomers can see, in the cloud decks, the temperatures drop to –236°F (–149°C) on Jupiter and –288°F (–178°C) on Saturn. But at great depths, the squeeze is on. By the time you reach 6,200 miles (10,000 kilometers) below the clouds on Jupiter, the pressure soars to 1 million times the barometric pressure at sea level on Earth. And the temperature equals that of the visible surface of the sun! But Jupiter is weirder than the sun. The density of the thick gas at this depth is much higher than at the solar surface, and the hot hydrogen gas is compressed so that it behaves like a liquid metal. Swirling currents of this liquid metal hydrogen generate powerful magnetic fields on Jupiter and Saturn that reach far out into space.

Jupiter and Saturn glow intensely in infrared light, each generating almost as much energy as it gets from the sun. (Earth, on the other hand, derives almost all its energy from the sun.) The upward moving heat, together with heat from the downward shining rays of the sun, stirs up their atmospheres and produces jet streams, hurricanes, and other kinds of atmospheric storms that continually change the appearance of these planets.

Almost a Star: Gazing at Jupiter

Jupiter has about 1,000th of the mass of the sun. Sometimes scientists call it "the star that failed." If only it had 80 or 90 times more mass, the temperature and pressure at its center would be so high that nuclear fusion would begin and keep going. Jupiter would start shining with its own light, making it a star!

Jupiter has a diameter of about 88,700 miles (143,000 kilometers). The gas-giant rotates at enormous speed, making one complete turn in only 9 hours, 55 minutes, and 30 seconds. In fact, Jupiter turns so fast that the rotation makes it bulge at the equator and flatten at the poles. With a clear look in steady air, you can detect this *oblate* shape through your telescope.

The rapid spin helps produce ever-changing bands of clouds, parallel to Jupiter's equator. What you see through your telescope when you view Jupiter is really the top of the planet's clouds. Depending on the viewing conditions, the size and quality of your telescope, and circumstances on Jupiter itself, you may see from as few as one to as many as twenty cloud bands (see Figure 8-1).

Figure 8-1:
Jupiter and its spinning-induced bands of clouds.

Jupiter's darker bands of clouds are called belts; the lighter bands are zones. When you look through a telescope, Jupiter looks like a round disk. Right down the center of the disk is the Equatorial Zone, flanked by the North and South Equatorial Belts (NEB and SEB). In the SEB, you may see the Great Red Spot, often the most conspicuous feature on Jupiter. This atmospheric disturbance, sometimes compared to a great hurricane, has hovered in the Jupiter atmosphere for at least 120 years. In fact, the Great Red Spot may have been spotted as early as 1664, although if so, it faded away before reappearing in the 19th century.

Jupiter is easy to find, because, like Venus (see Chapter 6), it shines brighter than any star in the sky. (A small exception: When its orbit takes it to the far side of the sun, Jupiter may be slightly fainter than the brightest star, Sirius.) If you have a computer-controlled telescope that can point to the position of the planet, or if you know where to look with binoculars or the naked eye, you can sometimes see Jupiter in the daytime.

When you can spot Jupiter with ease, you're ready for slightly more detailed observations. I provide directions for spotting the planet's features and moons in the following sections.

Scanning for the Great Red Spot

The Great Red Spot, shown in Figure 8-2, is a storm as big as Earth and sometimes bigger in the South Equatorial Belt. Like most of Jupiter's features, it can change from day to day. Its color can grow paler or deeper. White clouds, which are big enough to see with some amateur telescopes, form near the spot and move along the SEB. Sometimes a cloud in the SEB or another belt seems to be drawn out across the planet, stretched mostly in longitude. A cloud with this linear shape is known as a *festoon,* and spotting this interesting display is indeed a festive occasion!

If you don't see the Great Red Spot at first, it may be in a pale condition, but more likely, it has rotated to the backside of Jupiter. You have to wait for Jupiter to rotate back around. If you take telescopic views of the features of Jupiter at intervals of an hour or two during the night, you can see the spot and smaller features move across the disk of the planet as Jupiter rotates.

In the early 1990s, one of Jupiter's belts seemed to disappear overnight. Later, it reappeared. If this happens again, an amateur astronomer may well be the person who spots it first, because amateurs are always enjoying the ever-changing spectacle of Jupiter's belts and spots.

Figure 8-2:
Jupiter's
Great Red
Spot makes
for stormy
viewing.

Courtesy of NASA

Jupiter's invisible accessories

Did you know? Jupiter has thin rings made up of rock particles, perhaps tiny fragments from numerous boulders in the rings. Unlike Saturn's famous rings, Jupiter's accessories are dark, making them invisible through amateur telescopes. In fact, the rings are hard to see through any telescope, except with the Hubble and with instruments that probes carry right up to Jupiter. Microscopic particles from the rings are lifted up into a thick halo interior to the rings, which surrounds and perhaps merges into the upper atmosphere of the planet.

You can find pictures of Jupiter's rings at NASA's Planetary Photojournal site, `photojournal.jpl.nasa.gov/index.html`. Just click on the image of Jupiter and follow the links.

Shooting for Galileo's moons

Whenever the seeing is good, your telescope will reveal structure in Jupiter's cloudtops — belts, zones, spots, and maybe more — and you may see one or more of the planet's four large moons: Io, Europa, Ganymede, and Callisto. (Check out a photo of Jupiter and these moons in the color section of this book.)

Jupiter's four prominent moons (it has at least 34 smaller known moons as of January 2005) are known as the Galilean moons, or Galilean satellites, named after Galileo, their discoverer. Each of the big four moons orbits almost exactly in the equatorial plane of Jupiter, so each moon is always right overhead somewhere on Jupiter's equator. Any telescope worth owning can spot the Galilean moons, and many people can even see two or three of them through a good pair of binoculars. However, Io, the innermost of the Galilean moons, is hard to spot through binoculars because it orbits very close to the bright planet.

You can't see enough detail on any of Jupiter's moons with your own telescope to figure out what their surfaces are like, but you can notice differences in their brightnesses and, with careful study, perhaps in their colors.

If you take a look at space probe images of the Galilean moons, you can see that each moon is a little world unto itself, with composition and landscape that give it individual character. (See the section "Timing your moon gaze," later in this chapter, for Web sites to check the photos out.)

For basic details on the four major moons, check out the following list:

> ✔ Callisto has a dark surface, marked by many white craters. The surface is probably dirty ice — a mixture of ice and rock. The impacts of asteroids, comets, and big meteoroids have exposed the underlying clean ice; hence, the white craters.

- ✔ Europa has a ridged terrain that looks like rafts of ice. The surface may be a frozen crust that tops off an ocean of water and slush, perhaps 90 miles (150 kilometers) deep. Europa is the only place in the solar system outside of Earth where scientists have strong evidence that liquid water is present. The existence of liquid water on Mars, beneath a layer of permafrost, is only a theory.

- ✔ At 3,274 miles in diameter (5,268 kilometers), Ganymede is the largest moon in the solar system (even larger than 3,033-mile-wide Mercury). Ganymede's blotchy surface consists of light and dark terrains, perhaps ice and rock, respectively. The most noticeable marking is Valhalla, a huge-ringed impact basin about as large as the continental United States (judging the size by the outermost ring-ridge).

- ✔ Io's surface is peppered with more than 80 active volcanoes. This moon is the only place other than Earth where we have definite evidence of ongoing volcanism.

Although you can't enjoy the kind of up-close and personal view achieved with sophisticated space equipment, you can observe some interesting aspects of these moons as they orbit around Jupiter through your telescope. I cover phenomena that may block your view of the moons — such as occultations, transits, and eclipses — in the following sections.

Recognizing moon movements

Io, Ganymede, Europa, and Callisto are always moving, changing their relative positions and appearing and disappearing as they revolve around Jupiter. Sometimes you can see them all and sometimes you can't. If you can't spot one of the moons, here are a few possible explanations:

- ✔ An *occultation* may be underway, which occurs when one of the moons passes behind the limb of Jupiter (the edge of the disk you see through your telescope).

- ✔ The moon may be in *eclipse,* which occurs when the moon moves into Jupiter's shadow. Because Earth is often well to the side of a straight line from the sun to Jupiter, Jupiter's shadow can extend well to its side as you see it from the Earth. When you see a moon in plain view off the limb of Jupiter, and it suddenly dims and disappears, it has moved into the planet's shadow.

- ✔ The moon may be *in transit* across the disk of Jupiter; at that time, the moon is particularly tough to see, because the moons are pale in color, making them hard to spot against the cloudy atmosphere of Jupiter. In fact, a moon in transit can be much more difficult to discern than its shadow.

You can also observe a *moon shadow,* which occurs when one of the moons is sunward of Jupiter and casts a shadow on the planet. The shadow is a black spot, much darker than any cloud feature, moving across the planet.

The moon that casts the spot may be in transit at the time, but this isn't always the case. When Earth is well off the sun-Jupiter line, we may see a moon off the limb of Jupiter casting a shadow on the planet.

Timing your moon gaze

Sky & Telescope magazine carries a monthly schedule of the occultations, eclipses, shadow events, and transits of the four Galilean moons. Both that publication and _Astronomy_ also print a monthly chart showing the positions of the four moons with respect to the disk of Jupiter night by night. (See Chapter 2 for more about astronomy magazines.) You can tell which moon is which by comparing what you see through the telescope with the chart.

Remember the following general rules as you watch Jupiter's moons:

 ✔ All four Galilean moons orbit around Jupiter in the same direction. When you see them on the near side, with respect to Earth, they move from east to west, and when they orbit on the far side, they travel from west to east.

 ✔ A transiting moon is moving westward and a moon about to be occulted or eclipsed is heading eastward (following the east-west geographic directions in the sky of Earth).

Under excellent viewing conditions, you can discern markings on Ganymede, the largest Galilean moon, with a 6-inch or larger telescope. (See Chapter 3 for information on telescopes.) But to see the details of the surface, you need an image from an interplanetary spacecraft that visited the Jupiter system.

Com(et)ing within striking distance

On rare occasions, a comet strikes Jupiter, which causes a temporary dark blotch that may last for months. Scientists didn't know this until July 1994, when large chunks of the broken comet Shoemaker-Levy 9 struck Jupiter. But astronomers have gone back through old records of the markings on Jupiter and have found some suspicious features that may have been created in the same way.

The odds are poor that you'll witness a comet striking Jupiter, but you can keep the possibility in mind. If you see any new dark blotch, make good notes. An Arizona-based Canadian amateur astronomer named David Levy achieved international fame after he helped discover the Jupiter-smashing comet Shoemaker-Levy 9. Thanks to his lucid accounts of this and other astronomical events, he now commands lucrative fees for appearances, articles, and books. His writing on celestial stars appears regularly in _Parade,_ alongside other writers' personality profiles of the Hollywood variety. You, too, may enjoy that kind of notoriety — just keep a close watch on solar system traffic!

The best images of Jupiter and its moons are from the Galileo and Voyager 1 and 2 space probes and from the Hubble Space Telescope:

- ✔ You can find the Galileo images at `galileo.jpl.nasa.gov/gallery/index.cfm`.

- ✔ The Voyager images, plus some others, are at NASA's Planetary Photojournal Web site, `photojournal.jpl.nasa.gov`. When you see the planets, click on the picture of Jupiter.

- ✔ For Hubble images, examine the collection at the Space Telescope Science Institute at `hubblesite.org/newscenter/newsdesk/archive/releases/image_category`.

Our Main Planetary Attraction: Setting Your Sights on Saturn

Saturn is the second largest planet in our solar system, with a diameter of about 75,000 miles (121,000 kilometers). Most people are familiar with Saturn because of its striking set of rings. For centuries, astronomers thought that Saturn was the only planet that had rings. Today, we know that rings encircle all four gas-giant planets: Jupiter, Saturn, Uranus, and Neptune. But most of the rings are too dim to see through telescopes from the ground. The great exception is Saturn.

According to many observers, Saturn is the most beautiful planet. Not only are its famous rings easily visible through almost any telescope, but you can also spot Saturn's giant moon, Titan. Although many astronomers find Saturn's rings to be the celestial sight that most impresses their non-astronomer friends, Titan is also a worthy attraction.

In the following sections, I provide information on observing Saturn's rings, storms, and moons. Be sure to check out images of Saturn in the color section, too.

Ringing around the planet

Saturn's rings are usually easy to see because they're large and composed of bright particles of ice — millions of little ice fragments, some larger ice balls, and maybe some pieces the size of boulders. You can enjoy the rings through a small telescope and make out their shadow on the disk of Saturn (see Figure 8-3). Under excellent viewing conditions, the *Cassini division* — a gap in the rings named for the person who first reported it — may also be discernable.

Figure 8-3: Fragments of ice and rock make up Saturn's rings.

Measuring more than 124,000 miles (200,000 kilometers) across, Saturn's rings are only yards (or meters) thick. Proportionately, the rings are like "a sheet of tissue paper spread across a football field," as Professor Joseph Burns of Cornell University once wrote. But even though the rings are proportionately as thin as facial tissue, you wouldn't want to blow your nose in them. Stuffing ice up your nostrils may chill you out more than sniffing glue, but I definitely don't recommend it.

Saturn spins once every 10 hours, 39 minutes, and 22 seconds and is even more oblate — flattened at the poles — than Jupiter. The rings tend to mislead the eye a little, however, so noticing Saturn's squashed shape can be tricky.

Sometimes you have a hard time discerning Saturn's rings through the same telescope that revealed them in splendor just a few months before. They can even seem to vanish in small telescopes when you see them nearly edge-on.

The rings are very large but also very thin. They keep a fixed orientation, pointing face-on at one direction in space. There's a time each year when the rings are more face-on than usual, as seen from Earth, and a time three months later when they come closer to edge-on than usual.

As Saturn goes around its own 30-year orbit, however, there are times when the rings are precisely edge-on and seem to vanish through small (or sometimes even large) telescopes. You can't see the rings when their edges face Earth because they're extremely thin. On those occasions, with a powerful telescope, you may see the rings projected as a dark line against the disk of Saturn. The rings last disappeared in 1996, so don't worry about missing a chance to see the rings until the next vanishing act in September 2009. (If you miss that one, you have to wait until March 2025.)

Stormchasing across Saturn

Saturn has belts and zones just like Jupiter (see the section "Almost a Star: Gazing at Jupiter" earlier in this chapter), but Saturn's have less contrast and are much harder to see. Look for them during times of good atmospheric conditions, when you can use a higher power eyepiece on your telescope to spot planetary details.

About once every 30 years, a big white cloud, or "great white storm," appears in the Northern Hemisphere of Saturn. High-speed winds spread the cloud out until it forms a thick, bright band all the way around the planet. After a few months, it may disappear. Sometimes amateur astronomers are the first to spot a new storm on Saturn. The last great white storm was in 1990, so you may have to wait a while to see another. In the meantime, keep an eye out for smaller white cloud spots that spread partway around the planet.

Monitoring a moon of major proportions

Titan, Saturn's largest moon, is bigger than the planet Mercury. Its diameter is 3,200 miles (5,150 kilometers). Some large moons have thin atmospheres, but Titan has a thick, hazy atmosphere, composed of nitrogen and trace gases such as methane. Titan's atmosphere is hard to see through, but in 2004, the NASA Cassini space probe started mapping Titan's surface in infrared light (good for penetrating haze) and radar (even better). It looks flat and streaky, with few if any craters.

On January 14, 2005, the European Space Agency's Huygens probe landed on Titan. Pictures from Huygens showed what looked like ravines on the moon's surface with a possible shoreline in the distance. Astronomers suspect that there may be lakes or oceans of liquid hydrocarbons, such as ethane, on Titan. These first Huygens findings didn't prove that theory, but the information is consistent with it.

Born moons and converts orbiting in harmony

Moons come in two varieties: regular and other. The regular moons all orbit in the equatorial plane of their planet, and they all orbit in the same direction that the planet spins on its axis. This direction is called *prograde.* The regular moons almost certainly formed in place around Jupiter and Saturn, from an equatorial disk of protoplanetary and protomoon material. So Jupiter and Saturn, together with their many moons, are like miniature solar systems centered on big planets rather than stars.

But some of the small moons are like Elsa, the lioness that was "born free." It orbits in the opposite direction. Opposite orbits are *retrograde,* and these orbits may also be tilted with respect to the equatorial planes of their planets. The retrograde-orbiting moons formed elsewhere in the solar system, perhaps as asteroids, and were captured by the gravity of Jupiter and Saturn.

Jupiter has 38 confirmed moons and Saturn has 34, as of early 2005. Each planet probably has quite a few more small ones, and astronomers keep finding more. Any number you find in a printed book may be obsolete by the time you read it. Sometimes astronomers announce moons but don't count them. The officials at the International Astronomical Union want to be sure that the discoveries are confirmed. You can check the latest information on the natural satellites of these and other planets at the NASA Solar System Dynamics site at `ssd.jpl.nasa.gov/sat_discovery.html`. The moons without names are provisional discoveries awaiting confirmation.

With a good small telescope, you may be able to see two other moons, Rhea and Dione, when they're near their largest elongations from the planet. You can find a monthly chart of the locations of all the moons with respect to the disk of Saturn in the publication *Sky & Telescope.* Use the charts to plan your observations to spot Titan and perhaps Rhea and Dione. The best times to see the moons are usually when they approach greatest elongation from Saturn (see Chapter 6 for more on elongation). As of January 2005, Saturn had 34 confirmed moons.

The Cassini probe is in the midst of a long tour of Saturn and its moons that should extend until at least 2008. You can see pictures and other data that Cassini sends back at the Cassini-Huygens Multimedia page, `saturn.jpl.nasa.gov/multimedia/index.cfm`, and at the Cassini Imaging Central Laboratory for Operations (CICLOPS) site, `ciclops.lpl.arizona.edu`.

Chapter 9

Far Out! Uranus, Neptune, Pluto, and Beyond

In This Chapter

▶ Understanding the rocky, watery, gassy planets Uranus and Neptune

▶ Questioning the nature of Pluto

▶ Envisioning the Kuiper Belt

▶ Observing the outer solar system

Although Mars and Venus are closer to Earth, and Jupiter and Saturn are the bright, showy planets, observing the outer planets has its own mystique and offers its own rewards. This chapter introduces you to our solar system's three outer planets — Uranus, Neptune, and Pluto — and their moons. I offer useful tips for viewing these far-out worlds, and I also share details about the Kuiper Belt.

Breaking the Ice with Uranus and Neptune

The following are the most important facts about Uranus (pronounced *yoo-RAN-us,* or more commonly *YOO-rin-us*) and Neptune:

✔ They have a similar size with similar chemical compositions.

✔ They're smaller and denser than Jupiter and Saturn.

✔ Each planet is the center of a miniature system of moons and rings.

✔ Each planet shows signs of a long-ago encounter with a large body.

The atmospheres of Uranus and Neptune, like those of Jupiter and Saturn (see Chapter 8), are mostly hydrogen and helium. But astronomers call Uranus and Neptune *icy planets* because their atmospheres surround cores of rock and water. The water is so deep inside Uranus and Neptune and under such high pressure that it's a hot liquid. But when the planets merged and coalesced from smaller bodies billions of years ago, the water was all frozen.

You can tell a bona fide planetary scientist from a layperson, because the scientist calls the hot water inside Uranus and Neptune "ice," whereas the civilian innocently calls the hot water "hot water." Scientists use technical jargon the way predatory animals use scent markings — to proclaim their exclusive territory.

Uranus has about 14.5 times the mass of Earth, and Neptune equals 17.2 Earths, but they appear nearly the same size. The lighter Uranus is a bit larger, measuring 31,770 miles (51,118 kilometers) across the equator. Neptune's equatorial diameter is 30,784 miles (49,532 kilometers).

One day on Uranus lasts about 17 hours and 14 minutes; a day on Neptune spans 16 hours and 7 minutes. So, like Jupiter and Saturn, these planets both rotate much faster than Earth. Uranus takes about 84 years to make one trip around the sun, and Neptune takes about 165 years.

I cover more interesting facts about each planet in the following sections. Be sure to check out photos of Uranus and Neptune in the color section.

Bull's-eye! Tilted Uranus and its features

The evidence that Uranus suffered a major collision or gravitational encounter is that the planet seems to have flipped on its side. Instead of the equator being roughly parallel to the plane of Uranus's orbit around the sun, it nearly forms a right angle with that plane so that, in terms of Earth's directions, its equator runs roughly north-south.

Sometimes the North Pole of Uranus points toward the sun and Earth, and sometimes the South Pole faces our way. For about one-quarter of Uranus's 84-year orbit around the sun, its North Pole faces roughly sunward; for about another quarter, the South Pole faces roughly sunward; and the rest of the time, the sun illuminates the whole range of latitudes from pole to pole. In 2007, the sun will appear overhead at the equator on Uranus. That year looks like a good time to go to the beach, if Uranus had a beach. On Earth, the sun is never high in the sky at the North or South Pole, but on Uranus it sometimes passes overhead at the poles.

As of early 2005, Uranus has 21 known moons and 6 other reported, unconfirmed moons. It also has a set of rings. The rings are made of very dark material, probably carbon-rich rock, like certain meteorites known as carbonaceous chondrites. The moons and rings of Uranus orbit in the plane of its equator, just as the Galilean moons orbit in the equatorial plane of Jupiter (see Chapter 8), so the rings and the orbits of the moons of Uranus are at nearly right angles to the plane of the planet's orbit around the sun.

Essentially, you can think of Uranus and its satellites as a big bull's-eye that sometimes faces Earth and sometimes doesn't. Some large object probably hit that bull's-eye long ago and tilted it from its natural position.

Against the grain: Neptune and its moon

Neptune isn't tilted from the natural order; its equator is parallel to the plane of its orbit around the sun, or nearly so. Its rings are very dark, like those of Uranus, and probably consist of carbon-bearing rock.

Neptune has eight known moons and five more awaiting confirmation, as of early 2005. Its largest moon, Triton (which is larger than Pluto), has a diameter of 1,684 miles (2,710 kilometers). Seen from north and above, Neptune, like all the planets in our solar system, revolves counterclockwise around the sun. Most moons revolve counterclockwise around their planets. But Triton, which resembles a cantaloupe in photos from Voyager 2, goes against the grain, traveling clockwise around Neptune. After mulling (or meloning) this over, scientists concluded that Neptune captured Triton early in the solar system's history, probably after a collision in its vicinity. Triton became a moon when it may otherwise have been a planet similar to Pluto.

Triton consists of ice and rock, so it seems more like Pluto (see the next section) than Uranus or Neptune. Its surface is shaped by *cryovolcanism,* meaning eruptions and flows of cold icy substances rather than hot, molten rock. Water ice, dry ice, frozen methane, frozen carbon monoxide, and even frozen nitrogen are all present on Triton. The moon doesn't have many impact craters, probably because they got sloshed full of ice over time.

Environmental groups say that excessive tourism endangers national parks, so consider a trip to Triton. Its landscape is just as bizarre, and maybe as beautiful, as Yellowstone's. But if you head for Triton, expect a "Winter Wonderland." The surface has cold surges rather than hot springs, and its geysers spew long plumes of frigid vapor rather than torrid jets of steam. Just bring a space suit and some warm booties.

Meeting Pluto, an Unusual Planet

Pluto is the most distant planet from the sun in our solar system (see Figure 9-1). It moves inside the orbit of Neptune every 248 years for a few decades at a time, but the last such inside move ended in early 1999. It won't happen again in the lifetime of anyone now inhabiting Earth, unless medical research makes major strides between now and the 23rd century.

Figure 9-1:
Pluto is
mysterious,
rocky,
and icy.

Courtesy of NASA

Pluto is so far away that scientists have little idea about its geography. Its elongated elliptical orbit brings it within about 29.7 A.U. or 2.8 billion miles (4.4 billion kilometers) of the sun and takes it as far out as 49.5 A.U. or 4.6 billion miles (7.4 billion kilometers) away.

Images from the Hubble Space Telescope show lighter and darker regions on Pluto, which may correspond to areas of fresh ice and old ice on the planet, respectively. But that's about all scientists can tell at this point. See them at hubblesite.org/newscenter/newsdesk/archive/releases/image_ category, along with an animation of the turning globe of Pluto. (You can also see an image of Pluto in the color section of this book.)

No space probe has ever visited Pluto, but astronomers eagerly await such a trip. If all goes well, NASA will launch the New Horizons space probe to Pluto in 2006. It should get to Pluto in 2015 and then head on out through the Kuiper Belt, a region beyond Neptune where small, icy bodies abound (see the section "Buckling Down to the Kuiper Belt," later in this chapter, for more).

The moon chip doesn't float far from the planet

Pluto, like Uranus, is tilted on its side. Its equator is tilted about 120 degrees from the plane of its orbit. Astronomers assume Pluto, like Uranus, suffered a major collision. In fact, some astronomers believe that its moon, Charon, is a chip off Pluto's block, created by an impact on Pluto — much as our moon is believed to have formed from a great impact on Earth (see Chapter 5).

Pluto takes 6 days, 9 hours, and 17 minutes to turn once on its axis, and Charon orbits once around the planet in exactly the same amount of time.

So the same hemispheres of Pluto and Charon always face each other. In the earth-moon system, one hemisphere of the moon always faces Earth, but not vice versa. Someone standing on the near side of the moon can see the whole planet over the course of an Earth day, but a person standing on Charon can never see more than half of Pluto.

Pluto and Charon are both icy, rocky worlds, and unlike Uranus and Neptune, their ice is real, not molten. With a surface temperature of –387°F (–233°C), almost everything freezes on Pluto. Water ice, methane ice, nitrogen ice, ammonia ice, and even frozen carbon monoxide are present on the surface. It exhausts me to think of it! Scientists have detected some, but not all, of these substances on Charon.

Pluto isn't uniformly arctic, however. Astronomers suspect that it has some "tropical oases" where the temperature gets all the way up to –351°F (–213°C).

The little planet with little respect

Pluto is 1,430 miles (2,300 kilometers) in diameter, making it the smallest planet. It doesn't even measure up to the four Galilean moons of Jupiter and the moons Titan of Saturn and Triton of Neptune. In fact, Pluto is less than twice as big as 780-mile-wide (1,250-kilometer-wide) Charon. So astronomers often call Pluto and Charon a double planet.

But every so often a naysayer casts stones at Pluto by suggesting that it shouldn't even count as a planet. In 1999, some astronomers made an attempt to designate it as asteroid No. 10,000. Some astronomers who thought Pluto should be downgraded from a planet called it merely the largest object in the Kuiper Belt (see the next section). But other astronomers and plain folk defeated this plan. They argued that Pluto is round like a planet (most asteroids, other than the largest ones, are irregularly shaped), has a large moon, and has been considered a planet from the time American observer Clyde Tombaugh discovered it in 1930. Even if astronomers change the definition of a planet, Pluto should be grandfathered in.

Buckling Down to the Kuiper Belt

Scientists estimate that about 100,000 icy bodies — called Kuiper Belt Objects (KBOs) — larger than 60 miles (100 kilometers) in diameter orbit between Neptune's orbit and a distance of 50 A.U. from the sun. They remain beyond the reach of backyard telescopes, unless your backyard is on Neptune or one of its moons. Astronomers David Jewitt and Jane Luu discovered the first KBO in 1992. Since then, hundreds more have been found.

Astronomers haven't thoroughly surveyed the region of the Kuiper Belt, and experts calculate that among the thousands of KBOs that astronomers have yet to discover and study, one or two may be as big as Pluto. They may also be dimmer than Pluto, if they have darker surfaces and/or are farther from the sun. A discovery of such a KBO would rekindle the controversy over whether to call Pluto a planet (see the previous section), because nowadays a KBO isn't considered a planet. (Pluto was considered a planet, however, even before its discovery, when Percival Lowell thought that a planet beyond Neptune was influencing the motion of Uranus, as I describe in Chapter 17.) KBOs may even be large comets. (The Kuiper Belt is regarded as a huge collection of comets; see Chapter 4.)

Some of the hundreds of known KBOs share three properties with Pluto:

- They have highly elliptical orbits.

- Their orbital planes are tilted by a significant angle with respect to the plane of Earth's orbit.

- They make two complete orbits around the sun in approximately the same time that Neptune takes to make three orbits (496 years for Pluto's two orbits and 491 years for Neptune's three). This effect is called a *resonance,* and it works to keep Pluto and Neptune from ever colliding or even coming close to each other, although their orbits cross.

Pluto is safe from disturbance by the powerful gravity of the much larger Neptune and so are the KBOs that share these three properties — called *Plutinos,* meaning little Plutos.

Other kinds of objects are orbiting beyond Neptune and Pluto, but like the KBOs, the objects can't be very massive because their gravitational effects on known objects would give them away. One such body, called Sedna, was discovered in March 2004 at a distance of 90 A.U. from the sun, well beyond the 50 A.U. distance where the Kuiper Belt seems to fade out. Sedna's size isn't accurately known, but it's probably much smaller than Pluto. Some astronomers believe Sedna is a member of the Oort Cloud, which I describe in Chapter 4. The only large planets beyond Neptune and Pluto are the planets of other stars (see Chapter 14).

You can find out more about KBOs on Professor David Jewitt's Kuiper Belt Web site at `www.ifa.hawaii.edu/faculty/jewitt/kb.html`.

Viewing the Outer Planets

With experience, you can locate the large outer planets Uranus and Neptune, but tiny Pluto may be beyond your visual reach. The first time you look for any of the distant planets, you should seek the aid of a more experienced amateur astronomer (see Chapter 2).

The annual *Observer's Handbook* of the Royal Astronomical Society of Canada (www.rasc.ca) carries maps of the planets' positions during the year. Astronomy magazines (see Chapter 2) carry similar maps.

Sighting Uranus

Uranus was discovered with a telescope, and it sometimes shines bright enough to be barely visible to the eye under excellent viewing conditions. Through your telescope, you can distinguish Uranus from a star thanks to

- ✔ Its small disk, a few seconds of arc in diameter (defined in Chapter 6)
- ✔ Its slow motion across the background of faint stars

The disk of Uranus has a pale green tint; you can make out the disk with a high-power eyepiece when viewing conditions are good. (Chapter 3 has more about telescopes.) You can detect the motion of Uranus by making a sketch of its relative position among the stars in the field of view. For this purpose, use a low-power eyepiece so the field of view is larger and more stars are visible. Look again in a few hours or on the following night and sketch again.

As of early 2005, Uranus has 21 known moons and 6 more awaiting confirmation. Although you can glimpse a few of the biggest moons with large amateur telescopes, they're better suited for study with powerful observatory telescopes. Uranus's dark rings are detectable with the Hubble Space Telescope and in images made with very large telescopes on Earth, but you can't see them with amateur instruments.

You can see the Hubble Space Telescope images of these bodies at hubble site.org/newscenter/newsdesk/archive/releases/image_category by clicking on the Uranus link. You can browse through images of Uranus and its moons from the Voyager 2 space probe at photojournal.jpl.nasa.gov, the Planetary Photojournal Web site, again by clicking on the Uranus link.

Distinguishing Neptune from a star

Neptune appears fainter in the sky than Uranus, but it gets as bright as 8th magnitude (Chapter 1 has more about magnitudes). If Uranus challenges your observing skills, you really have to take them to the next level for Neptune!

Neptune is about the same actual size as Uranus, but it orbits much farther away, so through a telescope its apparent disk is smaller. You may need a large amateur telescope to discern it from a star. If you become good at perceiving pale hues in dim objects seen through a telescope, you'll be able to tell that Neptune has a blue tint.

Because Neptune orbits farther from the sun than Uranus, it moves at a slower speed. The slower speed combined with a greater distance from Earth means that the angular rate of speed across the sky — in arc seconds per day (see Chapter 6) — is *usually* less for Neptune than Uranus. So you may have to wait another night or two to be sure that you've seen Neptune moving across the background stars.

I say "usually" because both Uranus and Neptune, like all the planets beyond Earth's orbit, show retrograde motion at times (see Chapter 6), so they seem to slow down and reverse direction now and then. If you happen to catch Uranus when it changes direction in the sky, its apparent motion is much slower than usual, and by comparison, Neptune may go full tilt at that time.

Neptune has eight known moons and five more pending confirmation, as of early 2005. The largest is Triton (see the section "Against the grain: Neptune and its moon," earlier in this chapter, for more about Triton). After you master locating Neptune, look for Triton with a telescope of 6 inches or more in diameter on a clear, dark night. It has a large orbit, ranging about 8 to 17 arc sec from Neptune (about four to eight Neptune diameters), so you may mistake Triton for a star. But by sketching Neptune and the faint "stars" around it on successive nights, you can deduce which "star" moves with Neptune across the starry background as it also moves around Neptune. It takes Triton almost six days to make one full orbit around the planet.

You can browse through images of Neptune and its moons from the Voyager 2 space probe at `photojournal.jpl.nasa.gov`, the Planetary Photojournal Web site, by clicking on the Neptune link. You can see Hubble Space Telescope images at `hubblesite.org/newscenter/newsdesk/archive/releases/image_category`.

Straining to see Pluto

Pluto is a much tougher viewing challenge than any other planet in the solar system. It orbits far away and is small. Typically, Pluto is 14th magnitude (see Chapter 1). And it's currently moving away from the sun and Earth and will continue to move away for many years as it traverses its 248-year orbit.

Skilled amateurs claim to have seen Pluto with 6-inch telescopes. I recommend that you use at least an 8-inch telescope.

Pluto's moon, Charon, orbits very close to Pluto and revolves around it in just 6 days, 9 hours, and 17 minutes. You can distinguish it only with the most powerful observatory telescopes.

Part III
Meeting Old Sol and Other Stars

The 5th Wave   By Rich Tennant

EARLY ATTEMPTS TO CALCULATE THE EXACT TEMPERATURE OF THE SUN

"How's this? Hot enough for you?"

In this part . . .

Part III introduces you to the stars. No, not the wealthy people in Hollywood — I'm talking about the sun and all the other stars in the Milky Way galaxy and beyond. You find out about the types of stars and their life cycles, from birth to death. When Jennifer Lopez and Ben Affleck are long forgotten, Alpha Centauri will still be shining. When you weary of current events, remember that the real stars are still there for you.

I also provide a chapter on black holes and quasars, and I simplify the subject matter so you don't get a migraine trying to understand it. However, the information about space and time distortions may twist your thoughts a bit.

Chapter 10

The Sun: Star of the Earth

In This Chapter

▶ Understanding the makeup and activities of the sun

▶ Trying out techniques for viewing the sun safely

▶ Watching sunspots and eclipses

▶ Surfing the Web for solar images

*A*lthough many people are attracted to astronomy by the beauty of a moonlit night and a starry sky, you need nothing more than a sunny day to experience the full impact of an astronomical object firsthand. The sun is the nearest star to Earth and, in fact, provides the energy that makes life possible.

The sun is so commonplace in day-to-day life that people take it for granted. You may worry about getting sunburned and the effects of ultraviolet rays on your skin, but you probably seldom think of the sun as a primary source of information about the nature of the universe. In fact, the sun is one of the most interesting and satisfying astronomical objects to study, whether with backyard telescopes or advanced observatories and instruments in space. The sun changes hour by hour and day by day. And you can show it off to the kids without keeping them up past their bedtimes!

But don't even think about looking at the sun, let alone showing it to a child or anyone else, without taking the proper precautions that I explain in this chapter. You don't want your view of the sun to cost you your eyesight. Safety in viewing should be your prime consideration; after you know how to protect your vision with the proper equipment and procedures, you can follow the sun not only on a daily basis, but also over the 11-year sunspot cycle I describe later in this chapter.

This chapter introduces you to the science of the sun, the sun's effects on Earth and on industry, and to safe solar observing. Get ready to look at the sun in a new way — safely, and with awe.

Surveying the Sunscape

The sun is a *star,* a hot ball of gas shining under its own power with energy from *nuclear fusion,* the process by which the nuclei of simple elements combine into more complex ones. The energy produced by fusion inside the sun powers not only the sun itself, but also much of the activity in the system of planets and planetary debris that surrounds the sun — the solar system of which Earth is a part (see Figure 10-1, which is not to scale).

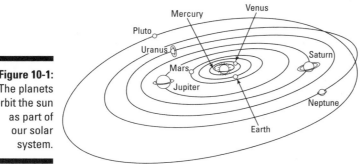

Figure 10-1:
The planets orbit the sun as part of our solar system.

The sun produces energy at an enormous rate, equivalent to the explosion of 92 billion one-megaton nuclear bombs every second. The energy comes from the consumption of fuel. If the sun consisted of burning coal, it would burn up every last lump of itself in just 4,600 years. But fossil evidence on Earth shows that the sun has been shining for more than 3 billion years, and astronomers are certain that it has been beaming away for longer than that. The estimated age of the sun is 4.6 billion years, and it still burns strong today.

Only nuclear fusion can produce the sun's huge total energy release — its *luminosity* — and keep it going for billions of years. Near the center of the sun, the enormous pressure and the central temperature of almost 16 million degrees Celsius (29 million degrees Fahrenheit) cause hydrogen atoms to fuse into helium, a process that releases the great torrent of energy that drives the sun.

About 700 million tons of hydrogen turn into helium every second near the center of the sun, and 5 million tons vanish and turn into pure energy.

If humans could generate energy through fusion on Earth, all the problems with fossil fuel, including air pollution and the consumption of nonrenewable resources, would be solved. But despite decades of research, scientists still can't do what the sun does naturally. Clearly, the sun deserves further study.

The sun's size and shape: A great bundle of gas

When I taught Astronomy 101, I'd always pose the question, "Why is the sun the size that it is?" Hundreds of mouths would drop open, dozens of pairs of eyes would wander the room, but hardly ever would anyone have a clue. It doesn't even seem like a logical question. Everything has a size, right? So what?

But if the sun is made of nothing but hot gas (and it is), what keeps it together? Why doesn't it all blow away, like a puffed-out smoke ring? The answer, my friend, is that gravity keeps the sun from blowing in the wind. Gravity is the force, which I describe in Chapter 1, that affects everything in the universe. The sun is so massive — 330,000 times the mass of planet Earth — that its powerful gravity can hold all the hot gas together.

Well, you may be wondering, if the sun's gravity pulls all its gas together, why isn't it squeezed down into a much smaller ball? The answer is the same thing that sells many a used car: high pressure. The hotter the gas, and the more gravity (or any other force) that squeezes it together, the higher its pressure. And gas pressure inflates the sun, just like air pressure inflates an automobile tire.

Gravity pulls in; pressure pushes out. At a certain diameter, the two opposite effects are equal and in balance, maintaining a uniform size. The certain diameter is 864,500 miles (1,391,000 kilometers), or about 109 times Earth's diameter. You could fit 1,300,000 Earths inside the sun, but I don't know where you'd get them.

The sun is round for much the same reason: Gravity pulls equally in all directions toward the center, and pressure pushes out equally in all directions. If the sun rotated rapidly, it would bulge a little at the equator and flatten slightly at the poles due to the effect that people often call *centrifugal force*. But the sun rotates at a very slow rate — only once every 25 days at the equator (and slower near the poles) — so any midriff bulge isn't noticeable.

The sun's regions: Caught between the core and the corona

The sun has two main regions on the inside and three on the outside (see Figure 10-2). The visible surface of the sun is called the *photosphere* (meaning "sphere of light"). The inside of the sun — in other words, the region below the photosphere — is called the *stellar interior*. At its center is the *core*. In the heart of the core, nuclear fusion generates all the sun's energy, and it releases it in the form of gamma rays, a very energetic type of light, and neutrinos, strange subatomic particles that I describe later in this chapter in the section "Solar CSI: The mystery of the missing solar neutrinos." The gamma rays bounce off one atom to another, back and forth, but on average move upward and outward. The neutrinos zip right through the whole sun and fly out into space. The farther out in the solar interior, the cooler the temperature gets.

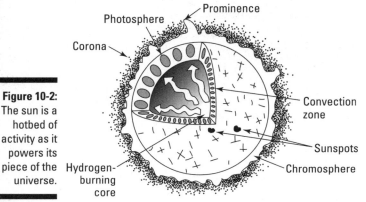

Figure 10-2: The sun is a hotbed of activity as it powers its piece of the universe.

The heart of the matter is that the sun's core is divided. The inner or *energy-generating* core extends to 108,000 miles from the center. Above that, the rest of the core is called the *radiative zone.*

At a distance of about 307,000 miles or 494,000 kilometers (about 71 percent of the way from the center to the visible surface, or photosphere), the core gives way to the next major region, the *convection zone.* In this region, huge streams of gas carry much of the upward-moving energy. The hot gas streams rise up, bringing heat along; they cool with altitude and then fall back down. The same type of process brings heat from the bottom of a kettle of boiling water up to the surface. Solar physicists believe that the magnetic fields of the sun, which cause sunspots and explosions of many kinds in the upper atmospheric regions, are generated mostly near the bottom of the convection zone.

The temperature of the convection zone drops from about 4 million°F (2.2 million°C) at its center to 9,900°F (5,500°C) when it reaches the next region, the photosphere. The photosphere is the layer of gas that produces all the visible light of the sun, except for what you can see at the time of a total eclipse (or with special instruments). The dark spots on the photosphere, or *sunspots,* are the most easily observable solar features.

If you look at the bright disk of the sun — do so only in accord with the safety instructions in the section "Don't Make a Blinding Mistake: Safe Techniques for Solar Viewing" later in this chapter — you see a portion of the photosphere.

Above the photosphere, the successive regions of the sun get hotter, not cooler. This fact remains one of the sun's great mysteries, and it has puzzled astronomers for decades. The *chromosphere,* or *color sphere,* is just above the photosphere. It only measures about 1,000 miles (1,600 kilometers) thick, but the temperature of the chromosphere reaches 18,000°F (10,000°C).

You can view the chromosphere at the edge of the sun if you use an expensive H-alpha filter, mentioned in the sidebar "Solar-viewing styles of the big spenders" later in this chapter, or you can see it in images taken with professional telescopes and displayed on the NASA and NOAA Web sites (see the section "Looking at solar pictures on the Net") and on various professional observatory Web sites. And you can see the chromosphere during a total eclipse of the sun, covered later in this chapter. During an eclipse, the chromosphere may appear as a thin red band all around the edge of the moon, which blocks the light from the photosphere.

Above the chromosphere is the corona, a region so rarified and electrified that the sun's magnetic field determines its shape. Where lines of magnetic force stretch and open outward into space, the coronal gas is thin and barely visible. It can readily escape in the form of solar wind (see the section "Solar wind: Playing with magnets"). Where lines of magnetic force reach up in the corona and then turn back down to the surface, they confine the coronal gas. The coronal region is thicker and brighter there. The corona is a sizzling 1.8 million°F (1 million°C) and in some places even hotter. Some loops extend from the photosphere up into the corona and contain gas much cooler than the surroundings. These loops are called *prominences,* which you can see on the limb of the sun during a total eclipse or at other times with an H-alpha filter.

The transition from the chromosphere to the hundred-times-hotter corona occurs in a very thin boundary layer called the *transition region.* It doesn't show up in views of the sun.

Solar activity: What's going on out there?

The term *solar activity* refers to all kinds of disturbances that take place on the sun from moment to moment and from one day to the next. All forms of solar activity, including the 11-year sunspot cycle and some even longer cycles, seem to involve magnetism. Deep inside the sun, a natural dynamo generates new magnetic fields all the time. The magnetic fields rise to the surface and on up to higher layers in the solar atmosphere where they twist around and cause all kinds of trouble. Recent observations show that additional magnetic fields are also generated in the higher layers of the atmosphere.

Astronomers measure magnetic fields on the sun by their effects on solar radiation, using instruments called *magnetographs.* You can see images taken with these devices on many of the professional solar observatory Web sites (see the section "Looking at solar pictures on the Net"). These magnetic field observations show that sunspots are areas of concentrated magnetic fields, and that sunspot groups have north and south magnetic poles. Outside of sunspots, the overall magnetic field of the sun is pretty weak.

Many of the rapidly changing features on the sun and probably all explosions and eruptions seem to be related to solar magnetism. Where there are changing magnetic fields, electrical currents occur (as in a generator), and when two magnetic fields bump into each other, a short circuit — called a *magnetic reconnection* — can suddenly release huge amounts of energy.

I cover several types of solar activity in the following sections.

Coronal mass ejections: The mother of solar flares

For decades astronomers believed that the main explosions on the sun were *solar flares.* They thought that solar flares occurred in the chromosphere (covered earlier in this chapter) and set things off.

Now astronomers know that they were just like the blind man who feels an elephant's tail and thinks that he knows all about the beast when he's touching one of the animal's least significant parts. Observations from space reveal that the primary engines of solar outbursts aren't solar flares, but *coronal mass ejections* — huge eruptions that occur high in the corona. Often, a coronal mass ejection triggers a solar flare beneath it in the low corona and chromosphere. You can see solar flares in many of the images on professional astronomy Web sites. As the number of sunspots increases over an 11-year sunspot cycle (see the following section), so does the number of flares.

Scientists didn't know about coronal mass ejections for many years because they couldn't see them. Astronomers could only get a good view of the corona at rare intervals during the brief duration of a total eclipse of the sun (see the section "Experiencing solar eclipses" later in this chapter). But solar flares can be seen at any time, so scientists studied them intensely and overestimated their importance.

Some of the prominences (see the previous section) that you can see on the edge of the sun with an H-alpha filter occasionally erupt. These eruptive prominences may also be phases of coronal mass ejections.

When satellite images show a coronal mass ejection that isn't going off, say to the east or to the west from the sun, but that forms a huge expanding ring or *halo event* around the sun, that's bad news. The halo event means that the coronal mass ejection — about a billion tons of hot, electrified, and magnetized gas — is heading right at Earth at about a million miles per hour. When it strikes the Earth's magnetosphere (which I describe in Chapter 5), dramatic effects sometimes result, as I describe later in the section "Solar wind: Playing with magnets."

If you see a halo event in one of the satellite images, check the National Oceanographic and Atmospheric Administration (NOAA) Space Environment Center Web site (www.sec.noaa.gov/today.html), because NOAA may be forecasting some pretty fierce space weather.

Cycles within cycles: The sun and its spots

Sunspots are regions in the photosphere where the magnetic field is strong and that appear as dark spots on the solar disk (see Figure 10-3). The spots are cooler than the surrounding atmosphere and often appear in groups.

The number of sunspots on the sun varies dramatically over a repeating cycle that lasts about 11 years — the famous *sunspot cycle*. In the past, people blamed everything from bad weather to a decline in the stock market on sunspots. Usually, 11 years pass between successive peaks (when the most spots occur) of the sunspot cycle, but this period can vary. Further, the number of spots at the peak can vary widely from one cycle to the next. No one knows why.

As a sunspot group moves across the solar disk due to the sun's rotation, the biggest spot on the forward side (the part of the group that leads the way across the disk) is called the *leading spot*. The biggest spot on the opposite end of the group is the *following spot*.

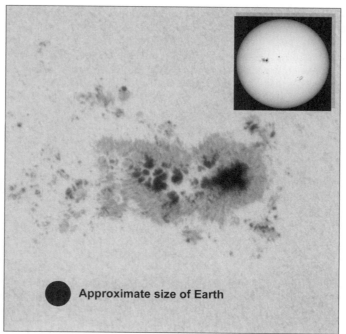

Figure 10-3:
A sunspot group 12 times bigger than Earth was photographed on September 23, 2000.

Courtesy of SOHO, NASA/ESA

Magnetograph observations show definite patterns in most sunspot groups. During one 11-year cycle, all the leading spots in the Northern Hemisphere of the sun have north magnetic polarity, and the following spots have south magnetic polarity. At the same time in the Southern Hemisphere, the leading spots have south polarity, and the following spots have north polarity.

Here's how these polarities are defined: The compass needle that points north on Earth is called a north-seeking compass. A north magnetic polarity on the sun is one that a north-seeking compass would point to. A south magnetic polarity on the sun is one that a north-seeking compass would point away from.

Just when you think that you have it straight, guess what? A new 11-year cycle begins, and the polarities all reverse. In the Northern Hemisphere, the leading spots have south polarity, and the following spots have north polarity. In the Southern Hemisphere, the magnetic polarities reverse, too. If you were a compass, you wouldn't know if you were coming or going.

To encompass all this information, astronomers have defined the sun's *magnetic cycle.* The cycle is about 22 years long and contains two sunspot cycles. Every 22 years, the whole pattern of changing magnetic fields on the sun repeats itself.

The solar "constant": Time to face the changes

The total amount of energy produced by the sun is called the *solar luminosity.* Of greater interest to astronomers is the amount of solar energy that Earth receives, or the *solar constant.* Defined as the amount of energy falling per second on 1 square centimeter of area facing the sun at the average distance of Earth, the solar constant amounts to 1,368 watts per square meter (127 watts per square foot).

Measurements made by solar and weather satellites sent up by NASA in the 1980s revealed very small changes in the solar constant as the sun turns. You may think that Earth receives less energy when dark sunspots are present on the solar disk, but that isn't the case; in fact, the opposite is true: more sunspots, more energy received from the sun. Chalk up another mystery for astronomers to solve.

According to astrophysical theory, the sun was slightly brighter when it was very young than it has been for the last several billion years, and it will cast more energy on Earth ages from now when it becomes a red giant star (see Chapter 11).

So "solar constant" sounds like wishful thinking, although from day to day and with amateur equipment, constant sounds pretty darn accurate.

Solar wind: Playing with magnets

Coronal mass ejections (covered earlier in this chapter) are usually invisible with amateur equipment but marvelously revealed by satellite telescopes. They spray billion-ton blobs of electrified gas, called *solar plasma,* permeated with magnetic fields, out into the solar system, where they sometimes collide with Earth's magnetosphere. (The *magnetosphere* is a huge region around Earth in which electrons, protons, and other electrically charged particles bounce back and forth from high northern latitudes to high southern latitudes, trapped in Earth's magnetic field. It acts as a protective umbrella against coronal mass ejections and the solar wind.)

A type of solar plasma called the *solar wind* is constantly streaming out from the solar corona. It moves through the solar system at about a million miles per hour (470 kilometers per second) as it passes Earth's orbit.

The solar wind comes in streams, fits, and puffs and constantly disturbs and replenishes Earth's magnetosphere, which becomes compressed in size and swells out again. The disturbances to the magnetosphere, especially those from traveling solar storms such as the coronal mass ejections, can cause displays of the Northern Lights (aurora borealis) and Southern Lights (aurora australis), as well as geomagnetic storms (see Chapter 5 for more on auroras). The geomagnetic storms can shut down power company utility grids (causing blackouts), blow out electronic circuits on oil and gas pipelines, interfere with radio communications, and damage expensive satellites. Some people even claim they can *hear* aurorae.

Solar disturbances and their effects on the magnetosphere are called *space weather.* You can see the latest official U.S. government space weather report and forecast at the Web site of the NOAA Space Environment Center (www. sec.noaa.gov/today.html).

Solar CSI: The mystery of the missing solar neutrinos

The nuclear fusion at the heart of the sun does more than change hydrogen into helium and release energy in gamma rays to heat the whole star. It also releases enormous numbers of *neutrinos,* or electrically neutral subatomic particles that have almost no mass, travel at nearly the speed of light, and can pass through almost anything.

A neutrino is like a hot knife through butter. It easily cuts through. In fact, neutrinos can fly right out from the center of the sun and into space. Those that head Earthward fly right through the earth and out the other side. But the neutrino is different from the hot knife, because the knife also melts butter that it comes in contact with. The neutrino just whooshes through without affecting the matter it passes through in almost (but not quite every) case.

The rare exceptions in which neutrinos do interact with matter can be detected in physics experiments, and therefore a tiny fraction of the solar neutrinos that pass through huge underground laboratories known as neutrino observatories do get counted. These observatories are located mostly in deep mines and tunnels under mountains. Deep down, few other kinds of particles fly around, so scientists have an easier time telling a solar neutrino from something else. One major facility, the Sudbury Neutrino Observatory in Canada, is 6,800 feet below the earth's surface. That's a good place to "delve deeply" into astronomy.

Counting neutrinos isn't easy, but some time ago reports from the neutrino observatories indicated a deficiency in solar neutrinos: The number of neutrinos coming to Earth was significantly fewer than the number scientists expected based on the rate at which the sun generates energy.

The solar neutrino deficiency was the least of our problems on Earth. It paled in significance beside AIDS, war, famine, the depletion of the forests, the extinction of valuable species, and the consumption of irreplaceable fossil fuel reserves. But the loss nagged at scientists, prompting them to make new theories of particle physics and to check on theoretical models of the solar interior.

Fortunately, scientists at the Sudbury Neutrino Observatory and elsewhere recently solved the problem of the missing neutrinos. It turns out that some of the neutrinos produced in the sun's core change to another neutrino type on the way to Earth, and earlier neutrino observatories, which reported the solar neutrino deficiency, couldn't detect the second type. The problem was a deficiency in laboratory equipment, not a deficiency in understanding how the sun generates energy or how many neutrinos it emits. Here's a good analogy: Suppose you have to count birds as part of an annual wildlife survey, but you wear eyeglasses with colored lenses. The colored glasses make it difficult to see birds of certain colors, so you may think that bluebirds are endangered when the problem is that you can see only cardinals.

Four billion and counting: The life expectancy of the sun

Some day, the sun must run out of fuel, so some day it will die. All good stars must come to an end. And without the energy and warmth of the sun, life on

Earth will cease to exist. The oceans would freeze, and so would the air. But what will actually happen is that the sun will swell up and take the form of a red giant star (see Chapter 11 for more about red giants). It will look enormous, and it will fry the oceans. So the oceans will actually evaporate before they have a chance to freeze.

Read the preceding paragraph carefully: I didn't say that the oceans *will* freeze; I said that they would freeze without the energy of the sun. In fact, the energy Earth receives will increase so much before the sun dies out that humans will die from the heat (if people still exist), not from the cold. And as for the seas, boiled tuna will be served, not frozen cod. Talk about global warming!

The red giant sun will puff off its outer layers, forming a beautiful, expanding nebula, the kind of shining gas cloud that astronomers call a planetary nebula. But no humans will be here to admire it. So to appreciate what we'll surely miss, take a good look at some of the planetary nebulas created by other stars (see Chapter 12).

The nebula will gradually fade away and all that will remain at its center is a tiny cinder of the sun, a hot little object called a *white dwarf star*. It won't be much larger than Earth, and although it will be very hot at first, it will be too small to cast much energy on Earth. Whatever's left on the surface of the earth will freeze. And the white dwarf will shine like an ember in a dying campfire, gradually fading away.

Fortunately, we should have about 5 billion years to go before that prospect looms near. Future generations can worry about this, along with the national debt and how to acquire rare first editions of *Astronomy For Dummies*.

Don't Make a Blinding Mistake: Safe Techniques for Solar Viewing

Seventeenth-century Italian astronomer Galileo Galilei made the first great telescopic discovery about the sun. By watching the daily movements of sunspots across the solar surface, he deduced that the sun rotates. By some accounts, he damaged his eyesight while researching the sun. Those stories may be wrong, but my warning isn't: Looking at the sun through a telescope or another optical aid such as binoculars is dangerous. A telescope or a pair of binoculars collects more light than the naked eye and focuses it on a small spot on your retina, where it can cause immediate, severe harm. Ever see a *burning glass,* a magnifying lens that focuses the rays of the sun on a piece of paper to set it on fire? Now you get the idea.

Looking at the sun with the naked eye isn't a good idea, and in some cases, it can be harmful. Taking even the briefest peek at the sun through a telescope, binoculars, or any other optical instrument (whether you use your property or someone else's) is very dangerous unless the device is equipped with a solar filter made by a reputable manufacturer specifically for viewing the sun. However, you can observe the sun with a technique called *projection* (see the following section). If you carefully follow the instructions in the next two sections, you shouldn't have a bad experience. But better yet, start your solar observing under the guidance of an experienced amateur or professional astronomer. (Head to Chapter 2 to find out about clubs and other resources that help you get going.)

Viewing the sun by projection

Galileo invented the *projection technique* by using a simple telescope to cast an image of the sun on a screen in the manner of a slide projector. This technique is safe only when used properly with simple telescopes, such as those sold under the description *Newtonian reflector* or *refractor.*

As I explain in Chapter 3, a Newtonian reflector uses only mirrors, aside from the eyepiece, and its eyepiece is near the top of the telescope tube, protruding at right angles. A refractor works with lenses and doesn't contain a mirror.

Don't use the projection technique with telescopes that incorporate lenses and mirrors along with eyepieces. In other words, don't use the projection technique with the Schmidt-Cassegrain or Maksutov-Cassegrain telescope models — including the highly regarded Meade ETX-90/PE telescope — which use both mirrors and lenses (I describe these telescopes in Chapter 3). The hot, focused solar image may damage the apparatus inside the sealed telescope tube and could pose a danger.

Here's how to safely view the sun with the projection technique:

1. **Mount a Newtonian reflector or refractor telescope on a tripod.**

2. **Install your lowest-power eyepiece in the telescope.**

3. **Point the telescope in the rough direction of the sun *without* sighting through or along the telescope; keep yourself and all other persons away from the eyepiece and not behind it where the focused solar beam emerges.**

4. **Find the shadow of the telescope tube on the ground.**

5. **Move the telescope up and down and back and forth while watching the shadow to make it as small as possible.**

The best way is for you or an assistant to hold a piece of cardboard beneath the telescope, perpendicular to the long dimension of the scope, so that the shadow of the tube falls on the cardboard. Move the telescope so that the tube shadow is as close to a solid, dark, circular shape as possible.

6. **Hold the cardboard at the eyepiece; the sun will be in the field of view and its image will project onto the cardboard.**

   If the sun's image isn't in view, the bright glare of the sun should be visible on one side of the cardboard; in that case, move the telescope to move the glare toward the center of the cardboard, thus moving the sun into view.

Figure 10-4 presents a diagram of the projection technique. The easiest and safest way to practice the technique is to consult an experienced observer from your local astronomy club; flip to Chapter 2 to find out how to locate a club in your area.

Even though you avoid looking through the telescope, you have to beware of some other hazards the projection method presents. I once saw a tough guy at a school in Brooklyn project a solar image with a 7-inch telescope. He didn't put his face near the eyepiece, but at one point, he moved his arm through the projected beam close to the eyepiece, where the solar image is very small. The hot concentrated image burnt a smoking hole in his black leather jacket.

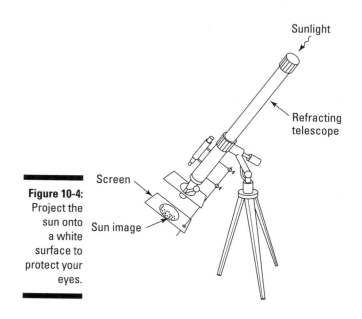

Figure 10-4:
Project the sun onto a white surface to protect your eyes.

Sunlight

Refracting telescope

Screen

Sun image

Solar-viewing styles of the big spenders

Special solar filters, called *H-alpha filters*, enable you to see many more of the sun's features than you can otherwise see in white light. In particular, filters are both necessary and wonderful for viewing solar prominences, which look like fiery arches on the edge or limb of the solar disk. But these filters are very expensive (typically over $1,000).

If the price doesn't deter you, and you have some experience in white-light solar observing as I describe in this chapter, you should investigate H-alpha filters. Two manufacturers that sell filters are Thousand Oaks Optical (www.thousandoaksoptical.com) and the Coronado Technology Group in Tucson, Arizona (www.coronadofilters.com). You may need a telescope adapter plate to hold your H-alpha filter, because the manufacturers don't necessarily make the filters to mate easily with specific telescope models.

Perhaps the safest and least expensive way to attain your own top-notch H-alpha-filtered views of the sun is to use a small telescope built for that sole purpose. Specifically, the well-regarded Personal Solar Telescope from Coronado gives outstanding views and costs only about $500. And the company now offers H-alpha binoculars that should work well too. But you can't use the Personal Solar Telescope or the H-alpha binoculars to look at anything but the sun.

You must take great care when using a telescope as a solar image projector, and you must *never* allow unsupervised children or any person who isn't trained in this method to operate the telescope. Don't look at the sun through your telescope, and don't look at the sun through the small finder telescope or viewfinder that your telescope may be equipped with. To avoid injury, make sure that no part of anyone's body, clothing, or other property gets in the projected beam of sunlight; only your cardboard projection screen should be in the beam.

When you get a good handle on the projection technique, you can look for sunspots. If you (pardon the expression) spot some spots, look again tomorrow and the next day to monitor their movement across the solar disk. In reality, although they may move a bit on their own, most of the sunspots' movements are due to the turning of the sun, or the *solar rotation*. You're repeating Galileo's discovery in a safe manner.

If you don't want to use the projection technique, or you have the kind of telescope that uses both lenses and mirrors — which you shouldn't use with this technique — you can still view the sun safely, but you need a special solar filter. Viewing the sun through a filter (see the following section) requires a significant investment, but the price you pay is well worth the viewing and the safety. Read on, Galileo!

Put the stop down on your telescope

When you block some or most of the light path of a telescope (for example, by using a filter that admits light only across part of the aperture), you *stop down* the telescope. Tell friends at the astronomy club that you view the sun with your telescope stopped down to see if they ask where the telescope was going before it stopped.

Guess who invented the stopping down of telescopes? Galileo! What a guy! You can repeat his work by watching sunspots with a stopped-down telescope. But he also conducted physics experiments, such as dropping weights from the Leaning Tower of Pisa. Don't even think about repeating that one. Stop before you drop.

Viewing the sun through front-end filters

The only solar filters that I recommend go at the *front end* of your telescope so that no light can enter the telescope without passing through the filter.

Filters at, near, or in place of the eyepiece may break as a result of concentrated solar heat, causing possible great damage to your eye. Use only filters that go at the front end of your telescope.

I give you the scoop on various types of scopes in Chapter 3, and I recommend the following front-end telescope filters for solar-viewing use:

- **Full-aperture filters:** Appropriate for telescopes of 4-inch aperture or less (the *aperture* is the diameter of the light-collecting mirror or lens in your telescope), such as the Meade ETX-90/PE (see Chapter 3). The filter extends across the full diameter of the telescope so the entire light-collecting mirror or lens receives the filtered light from the sun.

- **Off-axis filters:** Best for telescopes of 4-inch aperture or more — but not for refractors. An off-axis filter is smaller than the aperture of the telescope, but it's mounted in a plate that covers the entire aperture. The sun is so bright that you don't need the whole aperture of the telescope to collect enough light for good solar viewing. A larger aperture potentially can give a sharper view, but in most viewing locations, blurring by Earth's atmosphere nullifies this advantage. The less unneeded sunlight that gets into your telescope, the safer you and the telescope will be.

You want an off-axis solar filter with most telescopes other than refractors, because nonrefractors usually have little mirrors or mechanical devices on-center inside the telescope tube, blocking the part of the light that comes down the center of the tube.

In the special case of a refractor of 4-inch aperture or more, your filter should go over the top end of the telescope and be smaller than the telescope aperture, but it should mount centrally in the plate that covers the telescope. The filter should mount on-center because, generally speaking, the central part of the primary or objective lens of the refractor (the big lens) may have better optical quality than the periphery of the lens.

You can obtain solar filters from a variety of sources. Here are two suppliers with a reputation for good quality:

- **Rainbow Symphony,** in Reseda, California, sells "Solar Viewing Film" made from optical grade, aluminized polyester in 11-x-12-inch sheets for about $35 each. You can wrap a sheet around the top (front) of your telescope or binoculars and secure it with strong tape. See the Rainbow Symphony Web site at www.rainbowsymphony.com.

- **Thousand Oaks Optical,** in Thousand Oaks, California, manufactures full-aperture and off-axis glass solar filters under the designation Type 2 Plus. The filters are good for viewing through your telescope.

  Astronomers use the Thousand Oaks Type 3 Plus filters to photograph the sun through telescopes, but the filters aren't dark enough for use in telescopic viewing of the sun.

  Thousand Oaks also sells film-type solar filters, like Rainbow Symphony, but these filters are based on a black plastic polymer film. The company calls the product Black Polymer (go figure). See the Thousand Oaks Web site at www.thousandoaksoptical.com.

Only use solar filters in accordance with the manufacturer's directions.

Fun with the Sun: Solar Observation

The sun is a fascinating, constantly changing ball of hot gases that offers plenty of viewing opportunities for the prudent observer. With the proper precautions (see the preceding sections), you can see for yourself. In addition to observing the sun yourself, you can visit Web sites that offer awe-inspiring, professionally produced pictures. And taking advantage of both provides you with the full solar-viewing experience. This section suggests some ways that you can personally enjoy old Sol.

Tracking sunspots

After you become confident in your ability to observe the sun safely with the projection method or by equipping your telescope with a safe solar filter, you can begin your observation by tracking sunspots, using the following plan:

✔ Observe the sun as often as possible. (Tell the boss you didn't over-sleep; you were counting sunspots, not counting sheep.)

✔ Note the sizes and positions of sunspots and groups of sunspots on the solar disk.

Some sunspots look just like tiny dark spots. If the spot is truly a tiny dark spot, even through a powerful observatory telescope, it's a *pore*. But if a sunspot is big enough, you can distinguish its different regions. The dark central portion is called the *umbra,* and the surrounding area that appears darker than the solar disk but lighter than the umbra is the *penumbra.*

✔ Chart the motion of the sunspots as the sun makes one complete turn — which takes 25 days (at the equator) to about 35 days (near the poles; yes, the sun turns at different rates at different latitudes, another one of its many mysterious and unexpected properties).

The Solar Section of the Association of Lunar and Planetary Observers offers forms for recording and reporting solar observations at its Web site, www.lpl.arizona.edu/~rhill/alpo/solar.html. (Click on the ALPOSS report form link.)

As you track sunspots, you may want to note how many you see in a day; that figure is called (guess what?) the *sunspot number.* You may even want to keep track of the sunspot number from year to year to see if you can measure the sunspot cycle yourself. In the following sections, I give you info on how to compute the sunspot number and where to find official numbers.

Figuring your personal sunspot number

Compute your own sunspot number for each day of observation, using the formula:

$$R = 10g + s$$

R is your personal sunspot number, *g* is the number of groups of sunspots you see on the sun, and *s* is the total number of sunspots you count, including the spots in groups. Sunspots usually appear isolated from each other on different parts of the solar disk. Spots close together on one part of the disk are a group. And a completely isolated spot counts as its own group (the reasoning behind this designation is pretty spotty, but that's the way scientists have done it for years).

Suppose that you count five sunspots; three are close together in one place on the sun, and the other two appear at two widely separated locations. You have three groups (the group of three and the two groups consisting each of one spot), so *g* is three. And the number of individual spots is five, so *s* is five.

$$R = (10 \times 3) + 5$$
$$R = 30 + 5$$
$$R = 35$$

Finding official sunspot numbers

On the same day, different observers come up with different personal sunspot numbers. If you have better viewing conditions and a better telescope, or maybe just a better imagination, you calculate a higher sunspot number than your neighbor Jones. You calculate $R = 59$, and that bum Jones could only claim $R = 35$. When it comes to sunspot numbers, you're way ahead! Now when it comes to whose lawn is nicer, well, I'll leave that debate up to your neighbors. "R" you clear on this?

Central authorities, who tabulate and average together the reports from many different observatories, find by experience that some observers keep pace with Jones, some can't see as many, and some, like you, are far ahead. From this experience, the authorities calibrate each observatory or observer and make allowances in future counts so they can average the reports and get the best estimate of the sunspot number for each day.

You can check out the sunspot number every day (or whenever you want) at www.spaceweather.com.

Scientists expect the next peak of the sunspot cycle to occur around the year 2011. If you start watching spots before the peak, you may be able to judge the maximum for yourself, although scientists base the official decision on a complicated averaging scheme. And you can watch the number of spots decline over subsequent years until the sun approaches the minimum of the sunspot cycle, when you may look in vain for a decent sunspot for months at a time.

Experiencing solar eclipses

On a daily basis, the best way to see the sun's outermost, most varying, and most beautiful region, the corona, is to view the satellite images posted on the Web sites I list in the next section. But seeing the corona "live and in person" is a spectacle that you shouldn't deny yourself. The corona during an eclipse is one of nature's most beautiful sights. That's why many amateur astronomers save their earnings for years in order to splurge on a great eclipse trip (see Chapter 2 for details). And professional astronomers find ways to get to the eclipse, too, even though they have satellites and space telescopes at their disposal.

The sun experiences *partial, annular,* and *total eclipses* (see Figure 10-5). The greatest spectacle is the total eclipse; some annular eclipses are well worth the trip, too. (During an annular eclipse, a thin bright ring of the photosphere is visible around the edge of the moon.) A partial eclipse isn't something to drive hundreds of miles out of your way to see, because you don't see the chromosphere or corona, but you should definitely check it out if one comes your way. After all, the first and last stages of a total eclipse or an annular eclipse are partial eclipses, so you need to know how to observe those stages, too.

Figure 10-5:
What happens when the moon eclipses the sun.

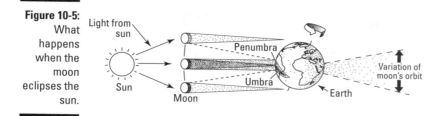

Light from sun

Penumbra

Sun

Umbra

Moon

Earth

Variation of moon's orbit

Observing an eclipse safely

To observe a partial eclipse, or the partial eclipse phases of a total eclipse, use the solar filters I describe in the section "Viewing the sun through front-end filters" earlier in this chapter. You can watch through binoculars or telescopes equipped with filters, you can hold a filter in front of your eyes, or you can use the technique I describe in the earlier section "Viewing the sun by projection."

A total eclipse normally starts with a partial phase, beginning with *first contact,* when the edge of the moon first comes across the edge of the sun. You now see a *partial eclipse* of the sun, signifying that you're in the *penumbra* or light outer shadow of the moon.

At *second contact,* the moon's leading edge reaches the far edge of the sun, totally blocking the sun from your view. Now you witness a *total eclipse;* you're in the dark *umbra* or central shadow of the moon. You can put down your viewing filter or your filtered binoculars and stare safely at the fantastic sight of the totally eclipsed sun. (For complete safety, you can check out an image of a total eclipse in the color section, too.) But you must not stare at the sun when totality is over.

The corona forms a bright white halo around the moon, perhaps with long streamers extending east and west. You may see thin bright polar rays off the north and south limbs of the moon and all around the lunar limb. (*Limb* is astronomy lingo that means "edge of the disk.") Watch for small, bright red points, which are solar prominences visible to the naked eye during brief moments of the eclipse. Near the peak of the 11-year sunspot cycle, the corona is often round, but near the sunspot minimum part of the cycle, the corona is elongated east-west. The corona takes a different shape during every eclipse.

Some people take the solar filters off their binoculars or telescopes and look at the totally eclipsed sun through these instruments without benefit of filters. This approach is very dangerous if

- You look too soon, before the sun is totally eclipsed

- You look too long (a very easy way to have an accident) and continue to watch through an optical instrument after the sun begins to emerge from behind the moon

---

Not your daddy's 3-D shades

Rainbow Symphony, a company I mention in the earlier section "Viewing the sun through front-end filters," is a prominent supplier of solar filters mounted in inexpensive eyeglass frames, like the 3-D viewers you use for certain movies. (The company sells 3-D viewers, too, but they don't do you any good at an eclipse.) Its product goes by the name "Eclipse Shades." These items are inexpensive, so I suggest bringing one pair of shades for each person in your party, even though you may purchase more expensive solar filters for your optical instruments. Usually, the organizers of tours and eclipse cruises distribute viewers, but sometimes they cut corners and hand out strips of aluminized Mylar. The substitute works, but the Mylar is less convenient than the mounted shades. See the Rainbow Symphony Web site at www.rainbowsymphony.com.

Be warned! I strongly advise against telescopic and binocular viewing of the sun without filters even during total eclipse, unless you're viewing under the direct control of an expert. Sometimes, for example, the experienced leader of an eclipse trip or cruise group uses a public address system, computer calculations, and personal observing know-how to announce when you can look at the eclipsed sun. The leader also tells you when you must stop with plenty of warning.

In my experience (which was painful), the easiest way to hurt yourself is to linger looking through your binoculars or telescope "just another second" when a tiny part of the bright visible surface of the sun starts to emerge from behind the moon. That tiny bright part may not make you immediately avert your vision, because it doesn't seem brilliant enough. But what you don't realize is that the infrared rays from the small, exposed part of the solar surface damage your eyes without dazzling them or causing immediate pain. In a few minutes or less, you begin to feel the pain. By then, the damage is done.

If you observe safely, follow all directions, and never take chances looking at the sun, you can look forward to many happy returns of total eclipses of the sun!

Seeking shadow bands and Baily's Beads

Another good reason to avoid looking at the sun with optical instruments during the total phase of an eclipse is that you have so much to look for all around the sky with your naked eye.

Here are some neat phenomena you can look for during an eclipse:

- ✔ Just before totality, so called *shadow bands,* shimmering low-contrast patterns of dark and light stripes, may race across the ground or, if you're at sea, across the deck of your ship. The stripes are an optical effect produced in Earth's atmosphere when the bright disk of the sun dwindles to the last little sliver behind the eclipsing moon before becoming completely eclipsed.

✔ *Baily's Beads* occur just instants before and after totality, when little regions of the bright solar surface shine through the mountains or crater rims on the edge of the moon.

✔ Wild animals (and domesticated animals if you're near a farm) react notably to the eclipse. Birds come down to roost, cows may head back for the barn, and so on. During one 19th-century eclipse, some top scientists set up their instruments in a barn, pointing the telescopes out through the door. Boy were they surprised when totality began and the livestock ran in!

When the sun becomes totally eclipsed, look at the dark sky all around the sun. You have a rare chance to see stars in the daytime. Special articles published in astronomy magazines or posted on their Web sites tell you which stars and planets to look for. Or you can figure it out yourself by simulating the date and time of the eclipse on your desktop planetarium, setting the program to display the sky as it will be from the place where you expect to observe.

Following the path of totality

Totality ends at *third contact,* when the moon's trailing edge moves out across the solar disk. At the last moment of totality, a small bright area of the photosphere may emerge from behind the moon. This stage is called the *diamond ring.* Now you're back in the penumbra, and you can see a partial eclipse. At *fourth* or *last contact,* the moon's trailing edge moves off the forward limb of the sun. The eclipse is over.

The whole eclipse, from first contact to last contact, may take a few hours, but the good part, totality, lasts from less than a minute to seven minutes or slightly more.

One place on the *path of totality* — the track of the center of the moon's shadow across the surface of Earth — boasts the greatest duration. Totality is briefer everywhere else on the path. Of course, the place where the eclipse has maximum duration may not be the place where the weather prospects are best, or it may not be a place you can easily or safely reach. So advanced planning of your eclipse trip is vital. At any good location, all the accommodations, rental vehicles, and so on are booked up at least a year or two in advance of the eclipse.

To plan your eclipse trip, pick a likely eclipse from Table 10-1 and start investigating the best way to view it.

Table 10-1	Future Total Eclipses of the Sun	
Date of Total Eclipse	**Maximum Duration (Minutes and Seconds)**	**Path of Totality**
Mar. 29, 2006	4:07	Eastern Brazil across the Atlantic to Ghana, across Africa to Libya, across the Mediterranean Sea to Turkey, across the Black Sea to Georgia and Kazakhstan
Aug. 1, 2008	2:27	Northern Canada, Greenland, Arctic Ocean, Russia, Mongolia, to China
July 22, 2009	6:39	India, Nepal, Bhutan, China, and across the China Sea and the central Pacific
July 11, 2010	5:20	Across the south Pacific and Easter Island to southern Chile and Argentina
Nov. 13, 2012	4:02	Australia and across the south Pacific toward (but not reaching) Chile
Nov. 3, 2013	1:40	Across the Atlantic to Africa, from Gabon to Uganda, Kenya, and Ethiopia
Mar. 20, 2015	2:47	Across the north Atlantic south of Greenland to the Faeroe Islands, Norwegian Sea, Spitsbergen Island, and on almost to the North Pole

A few years before each eclipse, articles with information on the weather prospects and logistics for viewing from various locations begin to appear in astronomy magazines. Check the *Sky & Telescope* and *Astronomy* magazine Web sites (see Chapter 2). Look for eclipse tour advertisements in the magazines and on the Web. Check out the most reliable eclipse predictions on the NASA eclipse site at sunearth.gsfc.nasa.gov/eclipse. See Chapter 2 for complete details on eclipse trips. And have a great time!

Looking at solar pictures on the Net

You can see current or recent professional photographs of the solar disk and sunspots (what solar astronomers call white-light photographs — white light being all the visible light of the sun) at various places on the Web. One good place to look is the site of Italy's Catania Astrophysical Observatory, web.ct. astro.it/sun. The white-light photo is the one labeled "Continuum," a technical term that means a colored filter wasn't used to take the picture. You can get experience in identifying a sunspot group and counting sunspots by practicing on these photos.

Sometimes the weather is cloudy in Italy, so you may need to look elsewhere for a professional white-light photo of the whole solar disk. A good place is the Web site of the Australian Space Weather Agency at www.ips.gov.au (click on the solar link to access the pictures).

When you become an advanced astronomer and want to photograph celestial scenes through your telescope, you may want to try solar photography. You can find inspiring examples at the Mount Wilson Observatory, where researchers have been photographing the sun since 1905. Check out the fantastic picture of an airplane silhouetted against a spotty sun and the picture of the largest sunspot group ever photographed, from April 7, 1947. Should you be lucky enough to see a sunspot group even half as large, you should be able to see it not only with a telescope, but also by looking through a solar filter without any other optical aid. The Mount Wilson site for white-light solar photographs is physics.usc.edu/solar/direct.html.

Astronomers study the sun in all kinds of light, not just white light. Their research includes pictures taken in ultraviolet and extreme ultraviolet radiation and X-rays, which are all forms of light invisible to the eye and, in fact, blocked by Earth's atmosphere. The pictures are made with telescopes mounted on satellites orbiting Earth at high altitude or taken by spacecraft located farther away and orbiting the sun just like Earth does. Sun images from satellites and from many kinds of telescopes on the ground are available on NASA's "Current solar images" Web site at umbra.nascom.nasa.gov/ images/latest.html.

If your computer is equipped to view movies over the Web, you can see selected videos of the changing face of the sun from the Solar and Heliospheric Observatory (SOHO) satellite (as long as it remains in operation) at NASA's SOHO Movie Theater site (sohowww.nascom.nasa.gov/data/synoptic/ soho_movie.html).

SOHO is a spacecraft built by the European Space Agency and loaded with scientific instruments, half of them supplied by NASA. So if you live in the United States or Western Europe, you probably paid taxes to support this project. And even if you're from some other nation or you don't pay taxes, you can still see the pictures.

Chapter 11

Taking a Trip to the Stars

In This Chapter

▶ Following the life cycles of the stars

▶ Sizing up stellar properties

▶ Checking out binary, multiple, and variable stars

▶ Meeting stellar personalities

▶ Stargazing for fun and for science

*H*undreds of billions of stars, like the sun, make up the Milky Way galaxy, where Earth resides. Likewise, the billions of other galaxies found in the universe all contain huge numbers of stars. And just like people, stars fit into dozens of classifications, and the overwhelming majority fall into several simple types. These types correspond to stages in the life cycles of stars, just as you classify people by their ages.

After you understand what a star is and how it runs through its life cycle, you get a feel for these shining beacons of the night sky, and those that aren't so bright, too.

In this chapter, I emphasize the initial mass of a star — the mass it's born with — as the main determinant of what the star will become. I continue with the key properties of stars, along with the features of binary, multiple, and variable stars that make them so interesting for you to observe.

And no discussion of stars is complete without some gossip about the celebrities, so I introduce you to some luminaries of the night sky that you should know — the leading "personalities" of the solar neighborhood.

Life Cycles of the Hot and Massive

The most important star categories correspond to successive stages in a star's life cycle: baby, adult, senior, and the dying. (What! No teenagers? The universe gave up on youth classifications after the terrible twos!) Of course, no astrophysicist worth a PhD uses such simple terms, so astronomers refer to the stages of stars as young stellar objects (YSOs), main sequence stars, red giants, and those in the end states of stellar evolution, respectively. (No star ever dies completely; at most, it "evolves" into a new and final state such as a white dwarf or a black hole.)

Here's the life cycle of a normal star with about the same mass as the sun:

1. **Gas and dust in a cool nebula condense, forming a young stellar object (YSO).**

2. **Shrinking, the YSO dispels its remaining birth cloud, and its hydrogen "fire" ignites.**

   In other words, nuclear fusion is underway, as I explain in Chapter 10.

3. **As the hydrogen burns steadily, the star joins the main sequence.**

   I describe this stage in stellar life in the section "Main sequence stars: A long adulthood" later in this chapter.

4. **When the star uses up all the hydrogen in its core, the hydrogen in the shell (a larger region surrounding the core) ignites.**

5. **The energy released by the burning of the hydrogen shell makes the star brighter and it expands, which makes its surface larger, cooler, and redder. The star has become a red giant.**

6. **Stellar winds blowing off the star gradually expel its outer layers, which form a planetary nebula around the remaining hot stellar core.**

7. **The nebula expands and dissipates into space, leaving just the hot core.**

8. **The core, now a white dwarf star, cools and fades forever.**

Stars with much higher masses than the sun have different life cycles; instead of producing planetary nebulae and dying as white dwarfs, they explode as supernovas and leave behind neutron stars or black holes. The life cycle of a massive star progresses rapidly; the sun may last 10 billion years, but a star that begins with 20 or 30 times the sun's mass explodes just a few million years after its birth.

Stars with lesser masses than the sun hardly have a life cycle at all. They begin as YSOs and join the main sequence to remain as red dwarfs forever. The explanation for this is a fundamental principle of stellar astrophysics: The bigger the mass, the fiercer and faster the nuclear fires burn; the smaller the mass, the less fiercely the fire burns and the longer it lasts.

By the time our sun uses up its core hydrogen, it will be at least 9 billion years old. But a red dwarf star burns hydrogen so slowly that it shines on the main sequence forever (for all practical purposes).

The following sections describe the stellar stages in more detail.

Young stellar objects: Taking baby steps

Young stellar objects (YSOs) are newborn stars that are still surrounded or trailed by wisps of their birth clouds. The classification includes *T Tauri stars,* named for the first of their type — the star "T" in the constellation Taurus — and *Herbig-Haro objects,* named for the two astronomers who classified them. (Actually, H-H objects are glowing blobs of gas expelled in opposite directions from the young star, which is usually hidden from view by dust from its birth cloud.) YSOs form in stellar nurseries — called *HII regions*— such as the Orion Nebula (see Figure 11-1), where hundreds of stars have been born in the past one or two million years.

Many of the Hubble Space Telescope images of spectacular jet-like nebulae are pictures of YSOs. The jets and other nebular surroundings are prominent, but the stars themselves are sometimes barely visible (if you can see them at all), hidden by the surrounding gas and dust. (Flip to Chapter 12 for more about nebulae.)

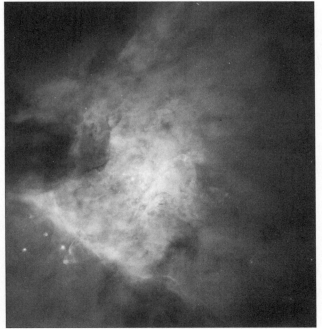

Figure 11-1:
The Orion Nebula cradles many young stellar objects.

Courtesy of C.R. O'Dell (Rice University) and NASA

Main sequence stars: A long adulthood

Main sequence stars, which include our sun, have shed their birth clouds and now shine thanks to the nuclear fusion of hydrogen into helium that goes on in the core (see Chapter 10 for more about nuclear fusion in the sun). For historical reasons, going back to when astronomers classified stars before they understood their differences, main sequence stars are also called dwarfs (never "dwarves"). A main sequence star is a dwarf even if it has 10 times more mass than the sun.

When astronomers and science writers refer to "normal stars," they often mean main sequence stars. When they write about "sun-like stars," they may mean main sequence stars with roughly the same mass as the sun, give or take a factor of no more than two. The writers may also be distinguishing stars on the main sequence, no matter how massive, from stars like white dwarfs and neutron stars.

The smallest main sequence stars — much less massive than the sun — are *red dwarfs,* which shine with a dull red glow. Red dwarfs have little mass, but they exist in massive quantities. The vast majority of main sequence stars are red dwarfs. Like tiny gnats at the seashore, they float all around you, but you can hardly see them. Red dwarfs are so dim that you can't see even the nearest one, Proxima Centauri — which is, in fact, the nearest-known star beyond the sun — without telescopic aid.

Red giants: Burning out the golden years

Red giant stars represent another kettle of starfish entirely. Red giants are much larger than the sun. Often, they measure as big around at the equator as the orbit of Venus or even the orbit of Earth. The giants represent a later stage in the life of an *intermediate-mass* star — one with several times more to somewhat less than the mass of the sun — after it graduates from the main sequence category (see the previous section).

A red giant doesn't burn hydrogen in its core; in fact, it burns hydrogen in a spherical region just outside the core, called a *hydrogen-burning shell.* A red giant can't burn hydrogen in its core because it has already turned its core hydrogen to helium through nuclear fusion.

Stars much more massive than the sun don't become red giants; they swell up so much that astronomers call them *red supergiants.* A typical red supergiant can be one or two thousand times larger than the sun and big enough to extend past the orbit of Jupiter, or even Saturn, if put in the sun's place.

The biggest stars are the loneliest

SETI observers (the Search for Extraterrestrial Intelligence — see Chapter 14) don't point their radio telescopes at massive stars to search for radio signals from advanced civilizations. Why not? Because massive stars explode and die after lifetimes so short that scientists can't imagine life evolving on any surrounding planets before the end comes.

Massive stars are much rarer than low-mass stars. The more massive the stars, the fewer they are. So eventually, as existing stars age and the birth clouds for new stars are used up, the Milky Way will consist overwhelmingly of just two types of stars: the red dwarfs that go on more or less forever and the white dwarfs that fade as they go. Yes, neutron stars and stellar mass black holes will dot the galaxy, but because they represent the remains of the much rarer, more massive stars, they'll be numerically insignificant compared to the red and white dwarfs, which come from the most abundant types of main sequence stars.

Stars are like people in that the biggest ones are rare, just as 7-foot, 5-inch basketball players are few and far between.

Closing time: Stars at the tail end of stellar evolution

The *end states of stellar evolution* is a catchall term for stars whose best years are far behind them. The category encompasses

- Central stars of planetary nebulae
- White dwarfs
- Supernovas
- Neutron stars
- Black holes

These objects are all stars on their final glide paths, doomed to oblivion.

Central stars of planetary nebulae

Central stars of planetary nebulae are little stars at the centers (no kidding) of small, beautiful nebulae. (You can see a photo in the color section of this book.)

Central stars of planetary nebulae are like white dwarfs and in fact turn into them. So the central stars, too, are the remains of sun-like stars. The nebulae, each composed of gas that a star expelled over tens of thousands of years, expand, fade, and blow away, and eventually they leave behind stars that no longer serve as the centers of anything — they become white dwarfs.

White dwarfs

White dwarfs can actually be white, yellow, or even red, depending on how hot they are. White dwarfs are the remains of sun-like stars that take after the old generals who, according to Douglas MacArthur, never die — they just fade away.

A white dwarf is like a glowing coal from a freshly extinguished fire. It doesn't burn any more, but it still gives off heat; it fades away over all eternity as it cools down. White dwarfs are the most common stars after red dwarfs, but even the closest white dwarf to Earth is too dim to see without a telescope.

White dwarfs are compact stars — small and very dense. A typical white dwarf may have as much mass as the sun, yet it takes up as much space as Earth, or a little bit more. So much matter is packed into such a small space that a teaspoon of a white dwarf would weigh about a ton on Earth. Don't try measuring it with your good silver; the spoon will get all bent out of shape.

Supernovas

Supernovas (which experts call *supernovae,* as though they all studied Latin like old-time scientists) are enormous explosions that destroy entire stars (see Figure 11-2).

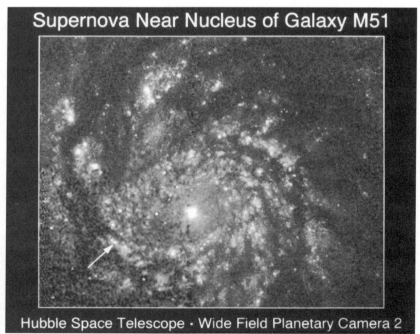

Figure 11-2:
A supernova in the spiral galaxy M51.

Supernova Near Nucleus of Galaxy M51

Hubble Space Telescope · Wide Field Planetary Camera 2

Courtesy of NASA

The first type you need to know about is Type II (hey, I didn't invent the numbering system). A *Type II supernova* is the brilliant, catastrophic explosion of a star much larger, brighter, and more massive than the sun. Before the star exploded, it was a red supergiant or maybe even hot enough to be a blue supergiant. Regardless of color, when a supergiant explodes, it may leave behind a little souvenir, which is a neutron star. Or much of the star may implode (fall in on its own center) so effectively that it leaves behind an even weirder object, a black hole.

The second type of supernova is called a Type Ia. *Type Ia supernovas* are even brighter than Type IIs, and they explode in a reliable manner. The actual brightness or luminosity of a Type Ia is always about the same; therefore, when astronomers observe a Type Ia supernova, we can figure out how far away it is by how bright it appears to us on Earth. The farther away it is, the dimmer the supernova looks. Astronomers use Type Ia supernovas to measure the universe and its expansion. In 1998, two groups of astronomers studying Type Ia supernovas discovered that the expansion of the universe isn't slowing down — it's expanding at an ever-faster rate. This discovery led experts to revise their theories of cosmology and the Big Bang (see Chapter 16).

Type Ia supernovas all produce similar explosions because they erupt in binary systems (covered later in this chapter) in which gas from one star flows down onto the other (a white dwarf), building up an outer hot layer that reaches a critical mass and then explodes, shattering the white dwarf. With less than critical mass, no explosion occurs. With critical mass, a standard explosion results. With more than the critical mass . . . wait — you can't have more than critical mass because the star already exploded! Astrophysics isn't so hard.

Neutron stars

Neutron stars are so small that they look up to white dwarfs, but they also outweigh them. (More accurately, they "outmass" them. Weight is just the force a planet or other body exerts on an object of a given mass. You would weigh different amounts on the moon, Mars, and Jupiter than you do on Earth, even though your mass stays the same — unless you overeat or stick to a crash diet.)

Neutron stars are like Napoleon: small in stature but not to be underestimated. (Figure 11-3 features a neutron star.) A typical neutron star spans only one or two dozen miles across, but it has half again or even twice the mass of the sun. A teaspoon of a neutron star would weigh about a billion tons on Earth.

Some neutron stars are better known as *pulsars*. A pulsar is a highly magnetized, rapidly spinning neutron star that produces one or more beams of radiation (which may consist of radio waves, X-rays, gamma rays, and/or visible light). As a beam sweeps past Earth like a searchlight from a galactic super-market opening, our telescopes receive brief spurts of radiation, which we call "pulses." So guess how pulsars get their name. Your pulse rate tells you how rapidly your heart is beating. A pulsar's rate tells how fast it spins. The rate can be a few hundred times per second or just once every few seconds.

Black holes

Black holes are objects so dense and compact that they make neutron stars and white dwarfs seem like cotton candy. So much matter is packed into such a small space in a black hole that its gravity is strong enough to prevent any-thing, even a ray of light, from escaping. Physicists theorize that the contents of a black hole have left our universe. If you fall into a black hole, you can kiss your universe good-bye.

You can't see the light from a black hole because the light can't get out, but scientists can detect black holes by their effects on surrounding objects. Matter in the vicinity of a black hole gets hot and rushes madly around, but it never gets organized; instead, the matter falls into the black hole and "that's all folks." All due to the powerful gravity of the black hole.

Figure 11-3:
A neutron star (at the arrow) photo-graphed by the Hubble Space Telescope.

Courtesy of Fred Walter (State University of New York at Stony Brook) and NASA

A view of Earth from the moon.

The moon, photographed by the Galileo spacecraft on its way to Jupiter.

Mercury, the planet closest to the sun, is frequently invisible to the naked eye, lost in the sun's glare.

Courtesy of NASA

Venus, covered in clouds, is the second brightest object in the night sky, after the moon.

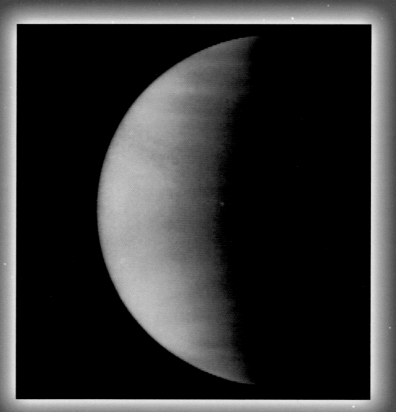

Courtesy of NASA

A panorama of the Columbia Hills on Mars, from the Mars Exploration Rover *Spirit*.

These sand dunes are in a crater on Mars.

Mars is likely to be the first planet visited by people from Earth.

Jupiter's four largest moons – Io, Europa, Ganymede, and Callisto – appear over the Great Red Spot in this montage.

Saturn's rings are easy to see.

Saturn's rings are shown here in false colors that reveal their nature. Those in turquoise are mostly ice; those in red contain many rock particles as well. The Cassini spacecraft photographed the rings in ultraviolet light, which is invisible to the human eye.

Like Saturn, Uranus has rings, but they can't be seen with a home telescope.

Courtesy of NASA

Streaky white clouds and a large dark spot marked Neptune's atmosphere when this picture was taken.

Courtesy of NASA

Clinging to planetary status, Pluto is about two-thirds the size of Earth's moon. Pluto's moon, Charon, is half the size of Pluto.

Courtesy of NASA

A total eclipse of the sun is one of nature's great spectacles.

Looking in infrared light that penetrates through dust clouds in the Milky Way, the Hubble Space Telescope photographed the Quintuplet Star Cluster near the center of our galaxy.

The globular cluster Messier 80 is a great ball of hundreds of thousands of stars.

Courtesy of NASA, The Hubble Heritage Team, STScI, AURA

The Eagle Nebula is a region in the Milky Way where new stars are being born.

Courtesy of NASA, Jeff Hester, and Paul Scowen, Arizona State University

The planetary nebula NGC 2392 in Gemini.

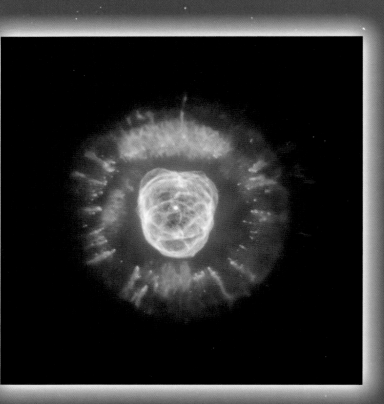

Two spiral galaxies are caught in a cosmic collision by the Hubble Space Telescope.

Actually, I oversimplified; a bit of the matter swirling around the black hole does escape — just in time, sometimes. The hole shoots it out in powerful jets at a significant fraction of the speed of light (which is 186,000 miles per second, in a vacuum such as outer space).

This is how scientists detect black holes:

- Gas swirls around that's just too hot for normal conditions.
- Jets of high-energy particles make their escape and avoid falling into the black hole.
- Stars race around orbits at fantastic speeds, driven by the gravitational pull of an enormous unseen mass.

Up until April 1999, when astronomers announced that they had discovered a third class of black holes — intermediate mass black holes — astronomers recognized two types of black holes:

- A *stellar mass black hole* has — you guessed it — the mass of a star. More accurately, these black holes range from about three times the mass of the sun up to perhaps a hundred times the solar mass, although astronomers haven't found any as massive as that. Stellar mass black holes are about the size of neutron stars. A black hole with 10 times the mass of the sun has a diameter of about 37 miles (60 kilometers). If you could squeeze the sun down to a size compact enough to make it a black hole (fortunately, this is probably impossible), its diameter would be 3.7 miles (6 kilometers). Stellar mass black holes form in supernova explosions and possibly by other means.

- A *supermassive black hole* has a mass of hundreds of thousands to even a few billion times the mass of the sun. Generally, supermassive black holes are located at the centers of galaxies. I want to say that they "gravitate" there, but most likely they form right there, or the galaxy forms around them. The Milky Way has a central black hole, known as Sagittarius A*. (Nope, the asterisk doesn't refer you to a footnote. You pronounce the name *Sagittarius A star.*) It weighs in at about 2.5 million solar masses, and we in the solar system orbit around that black hole once every 226 million years — the latest value from the Very Long Baseline Array, a radio telescope with component antennas that stretch across U.S. territory from the Virgin Islands, through North America, and out to Hawaii. Astronomers think that a supermassive black hole forms at the center of every galaxy, or at least the center of every full-sized galaxy. We're not so sure about dwarf galaxies. (Get the scoop on galaxies in Chapter 12.)

When I talk about the size of a black hole, I mean the diameter of its *event horizon.* The event horizon is the spherical surface around the black hole, where the velocity needed for something to escape the black hole equals the velocity of light. Outside the event horizon, the escape velocity is smaller, so light or even high-speed matter can escape.

Intermediate mass black holes have masses around 500 to 1,000 times the mass of the sun. They received their clever name from the experts who discovered them but weren't sure what they were. Some scientists think these holes are in the teenage stages of future supermassive black holes; holes that seem much lighter than they will be some day but that still swallow everything in sight. Others say that the holes may be something else entirely, but if so, what? Inquiring minds want to know, but right now more research is necessary to answer this question.

To tell the truth, supermassive black holes aren't stars. And, most likely, neither are intermediate mass black holes. But I have to mention them some place! You can't call yourself an astronomer if you don't know about black holes. (Check out Chapter 13 for even more about them.) After you feel ready to pass yourself off as an astronomer, folks will ask you all kinds of questions about black holes. But how many questions do you think you'll get about main sequence stars and young stellar objects?

Diagramming Star Color, Brightness, and Mass

The significance of the different types of stars (see the section "Life Cycles of the Hot and Massive") becomes clearer when you see basic observational data plotted on an astrophysicist's graph. The data are the magnitudes (or brightness) of the stars, which are plotted on the vertical axis, and the colors (or temperature), which are marked on the horizontal axis. This graph is called a *color-magnitude diagram,* or the Hertzsprung-Russell or H-R diagram, after the two astronomers who first made such diagrams (see Figure 11-4).

As an Astronomy 100 teacher, I could always tell who studied and who didn't. When I asked what data are plotted on the H-R diagram, the students who answered "H and R" revealed that they were just guessing.

Spectral types: What color is my star?

Hertzsprung and Russell didn't have good information on the colors or temperatures of stars, so they plotted the spectral type on the horizontal axis of their original diagrams. The *spectral type* is a parameter assigned to a star on the basis of its spectrum. The *spectrum* is the way that the light of a star appears when a prism or other optical device in an instrument called a spectrograph spreads it out.

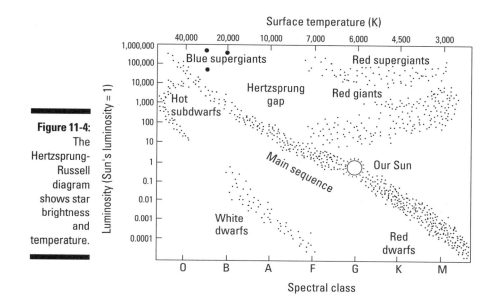

Surface temperature (K)

Figure 11-4:
The
Hertzsprung-
Russell
diagram
shows star
brightness
and
temperature.

At first, astronomers didn't have a clue about what spectral types meant, so they just grouped stars together (under the headings of Type A, Type B, and so on) based on similarities in their spectra. Later, some astronomers realized that the spectral types reflect the temperatures and other physical conditions in the atmospheres of the stars, where their light emerges into space. After scientists understood what the colors signified, they organized the spectral types in temperature order and Hertzsprung and Russell plotted them on their diagrams. Some of the original types were deemed superfluous and dropped.

The main spectral types on the H-R diagram are O, B, A, F, G, K, M, going from the hottest stars to the coolest. College students memorize this sequence with the help of mnemonic devices such as "Oh, be a fine girl (guy), kiss me." Table 11-1 describes the general properties of stars in each spectral class.

Table 11-1		Spectral Classes of Stars	
Class	*Color*	*Surface Temperature*	*Example*
O	Violet-white	30,000°K or more	Lambda Orionis
B	Blue-white	12,000°K to 30,000°K	Rigel
A	White	8,000°K to 12,000°K	Sirius
F	Yellow-white	6,000°K to 8,000°K	Procyon

(continued)

Table 11-1 *(continued)*

Class	Color	Surface Temperature	Example
G	Yellower-white	5,000°K to 6,000°K	Sun
K	Orange	3,000°K to 5,000°K	Arcturus
M	Red	Less than 3,000°K	Antares

Star light, star bright: Classifying luminosity

Each spectral class has subdivisions. For example, the sun has a G2V spectrum, which makes it

- A G-type star
- Slightly cooler than a G0 or G1 star
- Slightly hotter than a G3 star
- Much hotter than a K star
- A main sequence dwarf

You know the sun is a main sequence dwarf from the *V.* The *V* is called the sun's luminosity class. Each star has a luminosity class, represented by a Roman numeral.

Supergiants are luminosity classes I and II, giants are class III, and subgiants (a minor stage between main sequence and red giants) are luminosity class IV. All red dwarfs and other main sequence stars are luminosity class V, and white dwarfs are class D.

Today, you can find H-R diagrams that look quite different from each other in format but that all present the same data: the relative properties of the stars as indicated by their temperatures and brightnesses.

Some H-R diagrams are calibrated to show the true brightnesses or luminosities of the stars, not the apparent magnitudes or brightnesses as seen from Earth.

The brighter they burn, the bigger they swell: Mass determines class

A star with greater mass concentrates a fiercer nuclear fire in its core and produces more energy than a star of lower mass, so a more massive main sequence star is brighter and hotter than a less massive main sequence star. The more massive stars usually are bigger, too. With this information, you can follow the fundamental point of stellar astrophysics reflected in the H-R diagram: mass determines class.

On the H-R diagram (see Figure 11-4), magnitude (or luminosity) is plotted with greater luminosities or brighter magnitudes higher on the graph, and spectral class is plotted with the hotter stars to the left and the cooler stars to the right. Temperature runs from right to left, and magnitude runs from top to bottom.

On any H-R diagram, the plot of real observational data, where each point represents a single star, reveals a great deal to the careful reader:

✔ Most stars are on a band that runs diagonally from upper left to lower right. The diagonal band represents the main sequence, and all the stars on the band are normal stars like the sun, burning hydrogen in their cores.

✔ Some stars are on a wider, sparser, and roughly vertical band that stretches up and a little to the right from the diagonal band toward brighter magnitudes (higher luminosities) and cooler temperatures. This is the *giant sequence* and consists of red giant stars.

✔ A few stars are located all across the top of the diagram, from left to right. These are supergiants; blue supergiants are on the left side of the diagram, more or less, and red supergiants (which outnumber the blue) are on the right side.

✔ A few stars are located far below the diagonal band, down at the bottom left and bottom center of the diagram. These stars are white dwarfs.

Astronomers plot a main sequence star on the H-R diagram according to its brightness and temperature, but its brightness and temperature depend on only one thing: its mass. The diagonal shape of the main sequence represents a trend from high mass to low mass stars. The stars on the upper left of the main sequence have higher masses than the sun, and the stars on the far right have lower masses.

Astronomers usually don't plot young stellar objects on the same H-R diagram with all the other stars, but if they did, they would plot the YSOs on the right side of the diagram, above the main sequence — but not nearly as high up as the supergiant stars. Neutron stars and black holes are too dim to plot on the same H-R diagrams with normal stars.

Interpreting the H-R diagram

With just a little more explanation, you too can be a stellar astrophysicist and understand at one fell swoop why stars fall into different parts of the H-R diagram. Researchers spent decades figuring this out, but I give it to you on a plate. To keep it simple, I discuss a calibrated H-R diagram, where all the stars are plotted according to true brightness.

Consider this: Why is one star brighter or dimmer than another? Two simple factors determine the brightness of a star: temperature and surface area. The bigger the star, the more surface area it has, and every square inch (or square centimeter) of the surface produces light. The more square inches, the more light. But what about the amount of light a given square inch produces? Hot objects burn brighter than cool objects, so the hotter the star, the more light it generates per square inch of surface area.

Got all that? Here's how it all goes together:

- **White dwarfs** are near the bottom of the diagram because of their small size. With very few square inches of surface area (compared to stars like the sun), white dwarfs just don't shine as brightly. As they fade away like old generals, they move down the diagram (because they get dimmer) and farther to the right (because they get cooler). You don't see many white dwarfs on the right side of the H-R diagram because the cool stars are so faint that they usually fall below the bottom of the diagram as reproduced in books, and astronomers don't see and measure many faint dwarfs.

- **Supergiants** are near the top of the H-R diagram because they're super big. A red supergiant can be more than 1,000 times larger than the sun (if you put a supergiant in the sun's place, it would extend beyond Jupiter's orbit). With all that surface area, supergiants are naturally very bright.

  The fact that the supergiants are roughly at the same height on the diagram from left to right indicates that the blue supergiants (those on the left) are smaller than the red supergiants (the ones on the right). How do you know that? Supergiants are blue because they burn hotter, and if they burn hotter, they produce more light per square inch. Because their magnitudes are roughly the same (all the supergiants are near the top of the diagram), the red giants must have larger surface areas in order to produce equal total light (because they produce less light from each unit area).

- **Main sequence** stars are on the diagonal band that runs from the upper left to the lower right on the diagram, because the main sequence consists of all stars that burn hydrogen in their cores, regardless of their size. But a difference in size affects where the main sequence stars appear on the H-R diagram. The hotter main sequence stars (those on the left) are also bigger than the cool main sequence stars, so hot main sequence stars have two things going for them: They have larger surface areas, and they produce more light per square inch than the cool stars. The main sequence stars at the far right are the dim and cool red dwarfs.

The brown dwarfs don't top the charts

Brown dwarfs — discovered in the mid 1990s — are one of the latest additions to the celestial inventory. They're smaller than stars and about as big as a gas-giant planet like Jupiter, but much more massive. They shine by their own light like a star, not by reflected light, like Jupiter. But brown dwarfs aren't true stars because nuclear fusion only operates briefly in their cores. After fusion stops, they generate no more energy and just cool and fade. Their spectral types, L and T, signify objects cooler than type M and thus cooler than red dwarfs. On the H-R diagram in Figure 11-4, brown dwarfs would fall at the extreme bottom right or even just off the chart to the right of the corner.

Eternal Partners: Binary and Multiple Stars

Two stars or three or more stars that orbit around a common center of mass are called binary stars or multiple stars, respectively. Studies of binary and multiple stars help scientists to understand how stars evolve, and these small stellar systems are also fun to observe with backyard telescopes.

Binary stars and the Doppler Effect

About half of all stars come in pairs. These *binary stars* are *coeval,* a fancy term for "born together." Stars that form together, united by their mutual gravity as they condense from their birth clouds, usually stay together. What gravity unites, few celestial forces can break apart. A grown star in a binary system has never had another partner (well, hardly ever; some strange cases occur in dense star clusters, where stars can come so close that they may actually lose or gain a partner).

A binary system consists of two stars that each orbit a common center of mass. The center of mass of two stars that have exactly equal masses falls exactly halfway between them. But if one star has twice the mass of the other star, the center of mass is closer to the heavier star; in fact, the center is twice as far from the lighter star as from the heavier star. If one star has one-third the mass of its heavy companion, it orbits three times farther from the center of mass, and so on. The two stars are like kids on a seesaw: The heavier kid has to sit closer to the pivot so the two are in balance.

The two stars in a binary system follow orbits equal in size if the stars are equal in mass. Stars with different masses follow orbits of different sizes. The general rule: The big guys follow smaller orbits. You may think that binary systems are like our solar system, where the closer a planet orbits to the sun, the faster it goes and the less time it takes to make one complete orbit. That may be a reasonable idea, but it's wrong nonetheless.

In binary systems, the big star that follows the smaller orbit travels slower than the little star in a big orbit. In fact, their respective speeds depend on their respective masses. The star that carries one-third the mass of its companion moves three times as fast. By measuring orbital velocity, astronomers can determine the relative masses of a binary system's members.

The fact that the orbital speeds of the member stars of a binary system depend on mass is what makes binary stars attract the high interest of astronomers. There may be 40 ways to leave your lover, but astronomers have few ways to weigh the stars. Fortunately, instead of throwing up their hands in defeat, astronomers have been able to measure star masses by studying binary systems and taking account of the Doppler Effect.

If one star is three times more massive than the other, it moves around its orbit in the binary star system at one-third the orbital speed of its companion star. All astronomers have to do to figure out the stars' relative masses (meaning how much more massive one star is than the other) is measure their velocities. Only rarely can astronomers track the stars as they move, because most binary stars are so far away that we can't see them moving around their orbits. But even at great distances, we can receive the light from a binary star and study its spectrum — a spectrum that may be the combined light of both stars in the binary.

A phenomenon called the *Doppler Effect* helps astronomers figure out the masses of binary stars by studying their stellar spectra. Here's all you need to know about the Doppler Effect, named for Christian Doppler, a 19th-century Austrian physicist.

The frequency or wavelength of sound or light, as detected by the observer, changes, depending on the speed of the emitting source with respect to the observer. For sound, the emitting source may be a train whistle. For light, the source may be a star. (Higher frequency sounds have a higher pitch; a soprano has a higher pitch than a tenor.) Higher frequency light waves have shorter wavelengths, and lower frequency light waves have longer wavelengths. In the simple case of visible light, the shorter wavelengths are blue and the longest wavelengths are red.

According to the Doppler Effect:

- If the source is moving toward you, the frequency that you detect or measure gets higher, so

  - The pitch of the train whistle seems to be higher.

  - The light from the star seems to be bluer.

✔ If the source is moving away, the frequency gets lower, so

- The whistle you hear has a lower pitch.
- The star looks redder.

The train whistle is the official example of the Doppler Effect instructors have taught to generations of sometimes unwilling high school and college students. But who listens to train whistles anymore? A more familiar analogy is the way you feel the waves on the water as you zip around in a motorboat. As you ride toward the direction from which the waves are coming, you feel the boat rocked rapidly by choppy waves. But when you head back toward the beach, the rocking becomes much gentler and the same waves are less choppy. In the first case, you moved toward the waves, meeting them before they would've met you if you stood (or floated) still. So the frequency at which the waves struck the boat was greater than if the boat had remained at rest. The frequency of the waves doesn't change, but the frequency of the waves that you sense changes.

The spectrum of a star contains some dark lines — places (wavelengths or colors of light) where the star doesn't emit as much light as at adjacent wavelengths. The decreased emission at those wavelengths is caused by the absorption of light by particular kinds of atoms in the atmosphere of the star. The dark lines form recognizable patterns, and when the star is moving back and forth in its orbit, the Doppler Effect makes the patterns of lines move back and forth in the spectrum detected on Earth.

So by observing the spectra of binary stars and seeing how their spectral lines shift from red to blue to red again as the stars orbit, astronomers can determine their velocities and therefore their relative mass. And by seeing how long it takes for a spectral line to go as far to the red as it goes and then how long it takes to go as far to the blue as it goes and back to the red again, astronomers can tell the duration or period of the binary star orbit. When the spectral lines shift toward longer wavelengths, the phenomenon is a *redshift,* and when they shift toward shorter wavelengths, it's a *blueshift.* There are other ways of producing redshifts and blueshifts, but the Doppler Effect is the most familiar cause.

If you know that the period of one complete orbit is 60 days, for example, and if you know how fast the star is moving, you can figure out the circumference of the orbit and thus its radius. After all, if you drive nonstop from New York City to a town in upstate New York in three hours (good luck with the traffic!) at 60 miles per hour, you know that the distance you travel is 3 times 60, or 180 miles.

Stellar spectroscopy in a nutshell

Stellar spectroscopy is the analysis of the lines in the spectra of stars and by far an astronomer's most important tool for investigating the physical nature of stars. Spectroscopy reveals

- Radial velocities (motions toward or away from Earth) of stars

- Relative masses, orbital periods, and orbit sizes of stars in binary systems

- Surface gravities of stars

- Magnetic fields and their strengths on stars

- Chemical composition of stars (what atoms are present and in what states they exist)

- Sunspot cycles of stars (well, starspot cycles)

All this information comes from measuring the positions, widths, and strengths (how dark or how bright they are) of the little dark (or sometimes, bright) lines in the spectra of stars. Scientists analyze them with the help of the Doppler Effect to find out how fast stars move, the size of their orbits, and their relative masses. Other phenomena, such as the *Zeeman Effect* and the *Stark Effect,* affect the appearance of the spectral lines. Applying their knowledge of these effects, astronomers can figure the strength of the magnetic field of the star from the Zeeman Effect and the density and surface gravity in the star's atmosphere from the Stark Effect. The very presence of particular spectral lines, each of which comes from a specific kind of atom that's absorbing (dark lines) or emitting (bright lines) light in the atmosphere of a star, tells astronomers about some of the chemical elements present in the atmosphere of the star

and the temperatures in the star where those atoms are emitting or absorbing the light.

The spectral lines even tell astronomers what condition or *ionization state* the atoms are in. Stars are so hot that the heat may strip atoms of iron, for example, of one or more of their electrons, making them iron ions. Each type of iron ion, depending on how many electrons it has lost, produces spectral lines with different characteristic patterns and positions in the spectrum. By comparing the spectra of stars recorded with telescopes with the spectra of chemical elements and ions as measured in laboratory experiments or calculated with computers, astronomers can analyze a star without ever coming within light-years of it.

In cool stellar gases, much of the iron loses only one electron per atom, so it produces the spectrum of singly ionized iron. But in the very hot parts of stars, such as the million-degree corona of the sun, iron may lose 10 electrons; the element is in a high-ionization state, and it produces the corresponding pattern of spectral lines. That pattern clearly points to the existence of a very high temperature region on the star.

Certain parts of the sun's spectrum change along with the coming and going of disturbed regions on the sun, which peak about every 11 years. Similar changes occur in the spectra of other sun-like stars. So astronomers can tell the length of the sunspot cycle of a distant star by using spectroscopy, even though the star is too far away to ever catch a glimpse of its sunspots. (Well, not with present equipment, but that day will come.)

Two stars are binary, but three's a crowd: Multiple stars

Double stars are two stars that appear close to each other as seen from Earth. Some double stars are true binaries, orbiting their common centers of mass. But others are just *optical doubles,* or two stars that happen to be in nearly the same direction from Earth but at very different distances. They have no relation to each other; they haven't even been introduced.

Triple stars are three stars that appear to be in close proximity but, like the members of a double star, may or may not be all that close. But a *triple star system,* like a binary system, consists of three stars held together by their mutual gravitation that all orbit a common center of gravity.

A comparison to wedded (or unwedded) bliss may be in order. "Three's a crowd" is an expression of the instability in most romantic arrangements when a third person becomes involved. The same is true of triple star systems: They consist of a close pair or binary system and a third star in a much bigger orbit. If all three stars moved on close orbits, they would interact gravitationally in chaotic ways, and the group would break up with at least one star flying away, never to return. So a triple system is effectively a "binary star" where one member is actually a very close star-pair.

Quadruple stars are often "double-doubles," consisting of two close binary star systems that each revolve around the common center of mass of the four stars.

Multiple stars is the collective term for star systems larger than binaries: triples, quadruples, and more. At some point, the distinction between a large multiple star system and a small star cluster becomes blurred. One is essentially the same as the other.

Change Is Good: Variable Stars

Not every star is, as Shakespeare wrote, as "constant as the Northern Star." In fact, the North Star isn't constant either. The famous point of light is a *variable star,* meaning one whose brightness changes from time to time. For many years, astronomers thought that they had the North Star's brightness changes down pat. It seemed to brighten a little and fade a little over and over, reproducibly. But suddenly its expected changes, well, changed. This difference in the pattern may signify a physical change over time, and scientists are studying what it means. And recently, astronomers at Villanova University concluded that the North Star has brightened by about one magnitude (about 2.5 times) since antiquity.

Variable stars come in two basic types:

- *Intrinsic variable* stars change in brightness due to physical changes in the stars themselves. These stars divide into three main categories:
  - Pulsating stars
  - Flare stars
  - Exploding stars
- *Extrinsic variable stars* seem to change in brightness because something outside the star alters its light, as visible from Earth. The two main types of extrinsic variable stars are
  - Eclipsing binaries
  - Microlensing event stars

Going the distance: Pulsating stars

Pulsating stars bulge in and out, getting bigger and smaller, hotter and cooler, brighter and dimmer. These stars are in a physical condition where they simply oscillate like throbbing hearts in the sky.

Cepheid variable stars

The most important pulsating stars, from a scientific standpoint, are the Cepheid variable stars, named for the first studied star of their type, Delta in the constellation Cepheus (Delta Cephei).

American astronomer Henrietta Leavitt discovered that Cepheids have a *period-luminosity relation.* This term means that the longer the period of variation (the interval between successive peaks in brightness), the greater the true average brightness of the star. So an astronomer who measures the apparent magnitude of a Cepheid variable star as it changes over days and weeks, and who thereby determines the period of variability, can readily deduce the true brightness of the star.

Why do astronomers care? Well, knowing the true brightness enables us to determine the distance of the star. After all, the farther the star, the dimmer it looks, but it still has the same true brightness.

Distance dims stars according to the *inverse square law:* When a star is twice as far away, it looks four times as faint; when the distance is tripled, the star looks nine times as faint; and when a star is ten times farther away, it looks one hundred times as faint.

The headlines about the Hubble Space Telescope determining the distance scale and age of the universe came from a Hubble study of Cepheid variable stars. Those Cepheids are in faraway galaxies. By tracking their brightness changes and by using the period-luminosity relation, the Hubble observers figured out how far away the galaxies are.

RR Lyrae stars

RR Lyrae stars are similar to Cepheids but not as big and bright. Some are located in globular star clusters in our Milky Way, and they have a period-luminosity relation, too.

Globular clusters are huge balls of old stars that were born while the Milky Way was still forming. A few hundred thousand to a million or so stars are all packed in a region of space only 60 to 100 light-years across. Observing the changes in brightness of RR Lyrae stars enables astronomers to estimate their distances, and when the stars are in globular clusters, it indicates how far away the clusters are.

Why is it so important to know the distance of a star cluster? Here goes: All the stars in a single cluster were born from a common cloud at the same time, and they're all at nearly the same distance from Earth, because they exist in the same cluster. So when scientists plot the H-R diagram of stars in a cluster, the diagram is free of errors that may be caused by differences in the distances of the stars. And if scientists know the distance of the cluster, they can convert all the plotted magnitudes to actual luminosities, or the rates at which the stars produce energy per second. Such quantities can be directly compared with astrophysical theories of the stars and how they generate their energy. That's the stuff that keeps astrophysicists busy.

Long period variable stars

Astrophysicists celebrate the information gleaned from Cepheid and RR Lyrae variable stars. Amateur astronomers, on the other hand, delight in observing long period variables, also called Mira stars (or Mira variables). Mira is another name for the star Omicron Ceti, in the constellation Cetus (the Whale), the first known long period variable star.

Mira variables pulsate like Cepheids, but they have much longer periods, averaging 10 months or more, and the amount by which their visible light changes is even greater. At its brightest, Mira itself is visible with the naked eye, and at its faintest you need a telescope to spot it. The changes of a long period variable star are also much more variable than those of a Cepheid. The brightest magnitude that a particular star reaches can be quite different from one period to the next. Such changes are easily observed, and they constitute basic scientific information. You can help in these and other variable star studies, as I describe in the last section of this chapter.

Explosive neighbors: Flare stars

Flare stars are little red dwarfs that suffer big explosions, like ultrapowerful solar flares. You can't see most solar flares without the aid of special colored filters, because the light of the flare is just a tiny fraction of the total light of the sun. Only the rare, very large "white light" flares are visible without a special filter. (But you still need to use one of the techniques for safe solar viewing I describe in Chapter 10.) However, the explosions on flare stars are so bright that the magnitude of the star changes detectably. You're looking at the star through a telescope and suddenly it shines brighter. Not all red dwarfs have these frequent explosions. Proxima Centauri, the nearest star beyond our sun, is a flare star.

Nice to nova: Exploding stars

The explosions of novas and supernovas are so large that I can't lump them in with the flare stars; they're enormously more powerful and have much greater effects.

Novas

Novas explode through a build-up process on a white dwarf in a binary system, much like the Type Ia supernovas I describe earlier in this chapter, but the white dwarf in a nova isn't destroyed. It just blows its top and then it settles down, sucking more gas off its companion and onto its surface layer. The powerful gravity of the white dwarf compresses and heats this layer and after centuries or millennia, off it goes again! At least that's the theory. No scientist has been around long enough to see an ordinary or *classical nova* explode twice. But similar binary systems exist in which the explosions aren't quite as powerful as in a classical nova but in which the explosions recur frequently enough that amateur astronomers are always monitoring them, ready to announce the discovery of a new explosion and guide professionals to study it. These objects have various names, including *dwarf nova* and *AM Herculis systems.*

Classical novas, dwarf novas, and similar objects are known collectively as *cataclysmic variables.*

A nova bright enough to see with the naked eye occurs about once a decade, give or take. I studied one in Hercules for my doctoral thesis in 1963. If it hadn't exploded at just the right time, I could still be looking for a thesis topic. Most recently, the bright nova in Vela dazzled astronomers in 1999.

Supernovas

Supernovas throw off nebulae, called *supernova remnants,* at high speeds (see Figure 11-5). The nebula at first consists of the material that made up the shattered star, less any remaining central object, be it neutron star or black hole (see the section "Closing time: Stars at the tail end of stellar evolution" earlier in this chapter). But as it expands into space, the nebula sweeps up interstellar gas like a snowplow that accumulates snow. After a few thousand years, the supernova remnant is mostly swept-up gas rather than supernova debris.

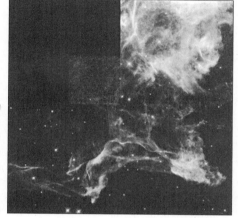

Figure 11-5:
A portion of the Cygnus Loop, a supernova remnant.

Courtesy of NASA

Supernovas are incredibly bright and rather rare. Astronomers estimate that, in a galaxy like the Milky Way, a supernova occurs every 25 to 100 years, but we haven't witnessed a supernova in our home galaxy since Kepler's Star in 1604, before the invention of the telescope. Others may have occurred, obscured from view by dust clouds in the galaxy. A huge southern star known as Eta Carinae looks like it may be on the verge of going supernova in the Milky Way, but in astronomers' parlance that means it may explode at any time — within the next million years.

Hypernovas

Hypernovas are especially bright supernovas that seem to produce at least some of the gamma ray bursts that flash in the sky from time to time. The bursts are extremely powerful blasts of high-energy radiation emitted in beams like the beams from searchlights. NASA launched the Swift satellite in November 2004 to discover more about them. When Swift detects a burst, it, well, swiftly notifies observatories on the ground to focus on the spot in the sky.

If you want to know more about what Swift discovers, go to NASA's Swift Web site at swift.gsfc.nasa.gov/docs/swift/swiftsc.html and click on <u>EDUCATION AND PUBLIC INFO</u>. All known hypernovas and gamma ray bursts have occurred in distant galaxies. That's just as well, because a gamma ray burst in our part of the Milky Way could have dangerous effects on the Earth if its beam struck the planet.

Stellar hide and seek: Eclipsing binary stars

Eclipsing binary stars are binary systems whose true brightness doesn't change (unless one of the two stars happens to be a pulsating star, flare star, or other intrinsic variable), even though they look like variable stars to us. The *orbital plane* of the system — the plane that contains the orbits of the two stars — is oriented so that it contains our line of sight to the binary system. That means that on every orbit, one star eclipses the other, as seen from Earth, and the star's brightness dips down during the eclipse. (And, of course, the situation reverses halfway through the orbital period when the eclipsed star now does the eclipsing.)

If the two stars in a binary system have orbital periods of four days, every four days the more massive star in the system, usually called *A,* passes exactly in front of the other star in the view from Earth. This blocks all or most of the light from star *B* from reaching Earth (depending on its size compared to *A* — sometimes the less massive star is larger than its heavy companion), and so the binary looks fainter. Astronomers call this event a *stellar eclipse.* Two days after the eclipse, star *B* passes in front of star *A,* creating another eclipse.

In the earlier section "Binary stars and the Doppler Effect," I mention how astronomers use the orbital velocities to figure out the relative masses of the stars. Well, they can also use the velocities to find the diameters of stars. Scientists take spectra and discover how fast the stars orbit by using the Doppler Effect, and they measure the durations of the eclipses in eclipsing binaries. A stellar eclipse of star *B* begins when the leading edge of *A* starts to pass in front of it. And the eclipse ends when the following edge of *A* finishes passing in front of *B.* So the orbital speed times the duration of the eclipse tells scientists how big star *A* is.

In all these methods, the fine details are a little more complicated, but you can easily understand the principles of stellar investigation.

The most famous eclipsing binary is Beta Persei, also known as Algol, the Demon Star. You won't have a devil of a time observing Algol's eclipses in the Northern Hemisphere — Algol is a bright star well placed for observation in the northern sky in the fall. You can watch its eclipses without a telescope or even binoculars. Every 2 days and 21 hours, Algol's brightness dims by a little over one magnitude — more than a factor of 2.5 — for about two hours. But you need to know *when* to look for an eclipse. You don't want to stand around in the backyard for almost three days. The neighbors will talk. Check out the pages of *Sky & Telescope*, where it lists information for observers. Usually, you find a paragraph called "Minima of Algol," which lists the dates and times when eclipses will occur for a period of a few months. (If you don't see a list in the current issue, it means that Algol has moved too close to the sun in the sky for observation that month.)

Minima are the times when variable stars, extrinsic or intrinsic, reach the lowest brightness levels of their current cycles. *Maxima* are the times when the stars shine brightest.

Hogging the starlight: Microlensing events

Sometimes a faraway star passes precisely in front of an even farther star. The two stars are unrelated and may be thousands of light-years from each other, but the gravity of the nearer star bends the paths taken by light rays from the farther star in such a way that the distant star appears much brighter from Earth for a few days or weeks. This effect is predicted from Einstein's Theory of General Relativity, and astronomers regularly detect it. When the gravity of a huge object like a galaxy bends light, astronomers call the process *gravitational lensing,* and when the gravity of a body as small as a star bends light, it's called *microlensing.*

You may be thinking how unlikely it is that two unrelated stars would line up perfectly with Earth, and you would be right! Congratulations on that thought. In order to detect such a rare event on a regular basis, astronomers use telescopic electronic cameras that can record hundreds of thousands to millions of stars at a time. With that many stars under observation, a foreground star will pass in front of a distant star every so often, even though astronomers don't know the stars in advance.

The trick is to point your telescope at a region where you can see vast numbers of stars simultaneously in the field of view. Such regions include the Large Magellanic Cloud, a nearby satellite galaxy of the Milky Way, and the central bulge of the Milky Way itself, where a whole mess of stars hang out.

Meeting Your Stellar Neighbors

You've already met Proxima Centauri, the nearest star beyond the sun (see the section "Main sequence stars: A long adulthood" earlier in this chapter). It orbits around a central point as the third or outlying member of the Alpha Centauri triple star system (see the section "Two stars are binary, but three's a crowd: Multiple stars" earlier in this chapter). For a complete look at the triple system, check out the following list:

✔ Alpha Centauri is a bright, G-type star in the southern constellation Centaurus (see Figure 11-6). It's a main sequence dwarf with about the same color as the sun but somewhat brighter.

✔ Alpha Centauri's orange companion is a slightly smaller and cooler dwarf named Alpha Centauri B.

✔ The little red dwarf and flare star Proxima is Alpha Centauri C.

Figure 11-6: Alpha Centauri is a triple star system in the far southern sky.

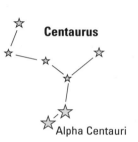

The Alpha Centauri system is about 4.4 light-years from Earth, with Proxima on the near side at 4.2 light-years.

Sirius, at a distance of 8.5 light-years, is the brightest star in the night sky. Its official name is Alpha Canis Majoris, in Canis Major (the Great Dog; see Figure 11-7). Slightly south of the Celestial Equator, Sirius is easily visible from most inhabited places on Earth. It's a white, A-type main sequence star that shines bright enough to make folks ask, "What's that big star?"

Figure 11-7: Sirius is top dog in Canis Major.

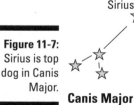

Like most stars other than the sun, Sirius has a companion: Sirius B, a white dwarf star. Sirius is known as the Dog Star, and when the American telescope-maker Alvan Clark discovered its tiny companion, Sirius B, in 1862, naturally someone named it "the Pup."

One legend and some written records that are open to differing interpretations suggest that Sirius was a red star a few thousand years ago. Despite much effort, astrophysicists have been unable to explain this color in terms of known physical processes, so naturally they say the story isn't true.

Vega is Alpha Lyrae, the brightest star in the constellation Lyra (the Lyre). It appears high in the sky at temperate northern latitudes (such as the U.S. mainland) on summer evenings and is an object that most amateur astronomers know like the back of their hands. Located about 26 light-years from Earth, Vega is a brilliant, white sparkler and one of the brightest stars in the sky.

Betelgeuse isn't really in the solar neighborhood; it's almost 500 light-years from Earth. But everybody likes its name, which many pronounce "Beetle Juice" (which is as good a way to say it as any), and observers enjoy its deep red color. It is, after all, a red supergiant about 50,000 times brighter than the sun. Although Betelgeuse is Alpha Orionis, the brightest star in Orion is actually Rigel (Beta Orionis).

Helping Scientists by Observing the Stars

Thousands of stars are under watch because they vary in brightness or exhibit some other special characteristic. Professional astronomers can't keep up with them all, and that's where you come in. You can monitor some stars with your own eyes, with binoculars, or with a telescope.

You need to be able to recognize the stars and to judge their magnitudes. The brightness of many stars changes so significantly — by a factor of two, ten, or even by hundreds — that eye estimates are sufficiently accurate to keep track of them. The trick is to use a *comparison chart,* a map of the sky that shows the position of the variable star and the positions and magnitudes of *comparison stars.* A comparison star has a known brightness that doesn't vary.

The American Association of Variable Star Observers (AAVSO) offers a wealth of information explaining how to observe variable stars. Its Web site is www.aavso.org. The association offers help of many kinds to observation novices. It also sells an inexpensive star atlas to help locate variable stars and offers a free set of CD-ROMs that contain over 4,500 charts of variable stars (you just pay for postage).

The AAVSO administers a Nova Search and a Supernova Search that you can join as you grow skilled in celestial observations:

✔ **Nova Search:** This program requires only patience, care, and a pair of 7 x 50 or 10 x 50 binoculars (see Chapter 3 for details on using binoculars and telescopes). When you join, you'll be assigned to concentrate on a small part of the sky. As frequently as possible on clear nights, you check your sky section, scanning it slowly with binoculars as you compare the pattern of stars against those on your charts.

If you find a "new star" (the original Latin meaning of *nova*) that doesn't appear on your charts, report it as fast as possible (preferably by e-mail). You may have discovered a real *nova,* an explosion in a particular kind of binary star. But you may want to wait a few hours to see if the "nova" moves. If it moves slightly with respect to other stars in the field of view, it isn't a star at all. It may be an asteroid or a faint comet.

Amateurs often make other kinds of innocent mistakes as well. For example, in the early 1950s (before e-mail), my pal Charlie and I sent a telegram to the AAVSO, announcing our discovery of a nova with a telescope on a rooftop in Brooklyn. We thought that it was a nova because it wasn't moving and it wasn't on the chart. But fame and fortune evaded us: We had only "discovered" a star that had been inadvertently left off the chart. It was all downhill for Charlie after that — he became a lawyer.

✔ **Supernova Search:** This program is for advanced amateurs. After a few years of stellar strivings, you should be ready for it. You need a decent telescope and preferably an electronic camera to make photographs through the telescope. Instead of monitoring a patch of sky in the Milky Way galaxy for nova explosions, you look at distant galaxies one by one, searching for a bright spot that may appear where you saw none the last time you looked. The bright spot could be a supernova. Because a supernova is so much brighter than a nova, you can easily see it in a distant galaxy.

Chapter 12

Galaxies: The Milky Way and Beyond

In This Chapter

▶ Tasting the Milky Way

▶ Sifting through star clusters

▶ Distinguishing different types of nebulae

▶ Classifying galaxies by shape and size

▶ Observing galaxies near and far

*O*ur solar system is a tiny part of the Milky Way galaxy, a great system of hundreds of billions of stars, thousands of nebulae, and hundreds of star clusters. The Milky Way, in turn, is one of the largest components of the Local Group of Galaxies. Beyond the Local Group is the Virgo Cluster, the nearest large cluster of galaxies — a full 50 million light-years from Earth. As scientists peer out into the universe over much greater distances, they see *superclusters,* immense systems that contain many individual clusters of galaxies. So far, we haven't found superclusters of superclusters, but *Great Walls,* which are immensely long superclusters, do exist. And much of the universe seems to consist of gigantic cosmic voids, which contain relatively few detectable galaxies.

This chapter introduces you to the Milky Way and its most important parts and takes you farther into the universe to meet other types of galaxies.

Unwrapping the Milky Way

Meet the Milky Way! A lot bigger than the candy bar, if not as sweet. But it does have a creamy-looking center. The Milky Way is the wide band of diffuse light that you can see best on clear summer and winter nights from a dark location.

Uncloaking the murky Milky Way

In the past, stargazers viewed the Milky Way with ease, but now many people can't see it or don't know it exists because they live in or near cities with so many lights that the sky is bright with light pollution rather than dark as nature intended.

The solution? Move away from the light pollution. Go out to the mountains or the shore on your vacation or on a weekend observing trip and check out a darker sky than you have at home. The light of the full Moon interferes with the Milky Way too, so plan your jaunt around the time of the new Moon, when there is little or no moonlight in the sky. The Milky Way is most prominent in the sky during summer and winter and least visible in spring and fall.

A stream of milk through the universe was as good as any explanation for the Milky Way until 1610, when Galileo observed it with a telescope. He found that the Milky Way is nothing to lap about: It consists of an immense number of dim stars that blend together into one large fuzzy region on the sky. Most of the stars in the Milky Way are invisible to the eye, but as a group they shine. Obviously, the telescope was a definite improvement for studying the Milky Way (and almost everything else in astronomy!).

Galaxies are the basic building blocks of the universe, and the Milky Way is a good-sized block. It contains almost everything in the sky that you can see with the naked eye — from Earth and our solar system to the stars of the solar neighborhood, the visible stars in the constellations, and all the stars that blend together to make that milky stream in the night sky — and plenty of objects and matter that you can't see. It also contains nearly every nebula you can see without telescopic aid and plenty more to boot.

Now that's a tall glass of milk! Besides loose stars, the Milky Way holds hundreds of star clusters, such as the Pleiades and Hyades in the constellation Taurus and, for lucky viewers in Australia, South America, and other points far south, the Jewel Box in Crux, the (Southern) Cross, and the magnificent star cluster Omega Centauri.

How and when did the Milky Way form?

The Milky Way is almost as old as the universe, and it certainly goes back more than the 12 billion years scientists estimate to be the age of some of its oldest stars. Long ago, gravity caused a gigantic cloud of primordial gas to fall together and condense. As little clumps inside the cloud collapsed even faster than the cloud as a whole, stars formed. Although the big cloud must have turned very slowly at first, it would've rotated faster and faster as it became smaller and flattened into the present day spiral disk structure. And before you knew it, *voilà, la voie lactee* (French for "there it is, the Milky Way"). Actually, its formation isn't quite that simple, because the Milky Way is a glutton — it has swallowed smaller neighbor galaxies for eons, adding their stars to its own collection, and it continues its feast today. What a menace!

The above is my favorite Milky Way theory, bar none. If you have a better one, become an astronomer yourself and write your own book someday — in science, theories make the world go 'round, and maybe even the galaxy.

What shape is the Milky Way?

The Milky Way is the shape it is and the size it is because, in the universe, gravity rules. The Milky Way is a spiral galaxy, consisting of a pizza-shaped formation of billions of stars (the *galactic disk,* about 100,000 light-years in diameter) that contains the spiral arms (see Figure 12-1). The arms are shaped roughly like the streams of water from a rotating lawn sprinkler and contain many bright, young blue and white stars and gas clouds. Groups of young, hot stars (called *associations*) dot the spiral arms in the galactic disk like pepperoni slices on a pizza. Bright and dark nebulae seem to mushroom all around the arms, and there are huge molecular clouds, such as Monoceros R2 (its location is marked in Figure 12-1), where most of the gas is very cool and dim. Between the arms are the *interarm regions* (not all astronomical terms are as catchy as Barnacle Bill, the name of a rock on Mars, or the Red Rectangle, a nebula shaped like an hourglass — go figure).

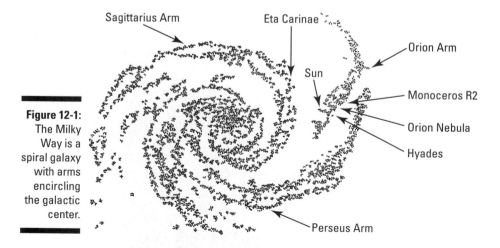

Figure 12-1:
The Milky Way is a spiral galaxy with arms encircling the galactic center.

At the center of our galaxy is a place called (you guessed it!) the *galactic center.* Centered on the center is the galactic bulge, which puts a sumo wrestler to shame. The *galactic bulge* is a roughly spherical formation of millions of mostly orange and red stars, sitting like a great meatball at the center of the galactic disk and extending far above and below it. And at the center of the bulge is Sagittarius A*, a supermassive black hole. Figure 12-1 presents a model of the Milky Way with its toppings and ingredients. (It's a close-up of the galactic disk, with the galactic bulge omitted for clarity.)

The flat imaginary surface or midplane of the galactic disk is called the *galactic plane,* and the circle that represents its intersection with the sky, as visible from Earth, is called the *galactic equator.*

Sometimes astronomers list the position of an object in galactic coordinates rather than in right ascension and declination (coordinates that I define in Chapter 1). The galactic coordinates are *Galactic Latitude,* which is measured in degrees north or south of the galactic equator, and *Galactic Longitude,* which is measured in degrees along the galactic equator.

Galactic Longitude starts at the direction to the galactic center, which is zero degrees longitude. (Actually, the zero point of Galactic Longitude is slightly off the galactic center, because scientists placed it where the galactic center was thought to be in 1959; we know better now.) Galactic Longitude proceeds along the galactic equator from the constellation Sagittarius into Aquila, Cygnus, and Cassiopeia; it goes on through Auriga, Canis Major, Carina, and Centaurus; and all the way to 360 degrees longitude, back at the galactic center. When you look with binoculars at the constellations I just named, you see more stars, star clusters, and nebulae than elsewhere in the sky. The "plane truth" is that the constellations that the galactic plane intersects are among the finest sights in the sky.

You can find panoramic maps of the Milky Way's galactic plane — as recorded with radio telescopes, X-ray and gamma-ray observatory satellites, and visible light (or "optical") telescopes on the ground — at NASA's MultiWavelength Milky Way Web site at `adc.gsfc.nasa.gov/mw/mmw_images.html`.

Where can you find the Milky Way?

The Milky Way isn't a certain distance from Earth; it contains our planet. The galactic center is about 25,000 light-years from Earth. Recent measurements with a radio telescope named the *Very Long Baseline Array* show that the solar system takes about 226 million years to make one orbit around the galactic center. The measurement cleared up a discrepancy: Scientists were unsure whether this interval, the *galactic year,* was 200 million years or 250 million years. Now they have an accurate number.

Seeing beyond the Milky Way

The three objects that are beyond the Milky Way but readily visible with the naked eye are the Large and Small Magellanic Clouds (two nearby galaxies visible from the Southern Hemisphere) and the Andromeda Galaxy. Some people blessed with excellent vision (and many others who want to impress their friends) say that they can see the Triangulum Galaxy too. Both the Andromeda and Triangulum Galaxies are about 2 million light-years from Earth, but Andromeda is bigger and brighter.

I count the Large Magellanic Cloud as one object, but it actually contains a huge bright nebula, the Tarantula, which you can make out with the naked eye. For a few months in 1997, you could see a bright supernova in the Large Cloud, Supernova 1987A. It was the first supernova visible to the naked eye since Kepler's Star in 1604, although it wasn't in our own galaxy like Kepler's Star. It was such a rare event that I flew to Chile to view it. The thrill was worth the long flight and the expense (which, fortunately, was paid by a magazine that awaited my breathless report).

The outskirt of the galactic disk, or *galactic rim* — as science fiction fans know it — is, at its closest part to Earth, about an equal distance in the opposite direction from Sagittarius. The disk of the Milky Way is pretty much identical with the milky band of light in the sky.

The Milky Way is about 169,000 light-years from a galaxy called the Large Magellanic Cloud, about 2.6 million light-years from the Andromeda Galaxy, and about 50 million light-years from the nearest big cluster of galaxies, the Virgo Cluster. It also falls smack dab in the middle of a little cluster of galaxies (sizes are relative here), the Local Group.

Star Clusters: Galactic Associates

Star clusters are bunches of stars located in and around a galaxy. They haven't associated by chance (even though one type of star cluster is called an *association*); they're groups of stars that formed together from a common cloud and, in most cases, are held together by gravity. The three main types of star clusters are *open clusters, globular clusters,* and *OB associations.*

For superb images of star clusters, consult the Anglo-Australian Observatory at www.aao.gov.au/images or treat yourself to a coffee-table book — *The Invisible Universe* by David Malin (Bulfinch Press, 1999) — that contains a collection of the greatest photographs from the observatory. You can pick up the book in French, German, Italian, and Japanese editions, because everyone loves these great pictures. You can also check out a photo of a star cluster in the color section of this book.

A loose fit: Open clusters

Open clusters contain dozens to thousands of stars, have no particular shape, and are located in the disk of the Milky Way. Typical open clusters span about 30 light-years. They aren't highly concentrated (if at all) toward their centers, unlike globular clusters (see the next section), and open clusters typically are much younger. Open clusters are great for viewing with small telescopes and binoculars, and you can see some with the naked eye. You can find them marked on most good star atlases, such as *Norton's Star Atlas* by Ian Ridpath (20th Edition, PI Press, 2003).

The most famous and easily seen open clusters in the northern sky are

- ✔ The Pleiades, in the northwest corner of Taurus, the Bull

  The Pleiades, also known as the Seven Sisters, looks with the naked eye like a tiny dipper. You can compare your eyesight with a friend's by seeing how many stars you can count in the Pleiades, which is M45, the 45th object in the *Messier Catalog* (see Chapter 1). Gaze through binoculars and see how many more you can find. The brightest star in the Pleiades is Eta Tauri (3rd magnitude), also called Alcyone. (See Chapter 1 if you need an explanation of magnitude.)

- ✔ The Hyades, in Taurus

  The Hyades, also a great sight for the naked eye, includes most of the stars that make up the "V" shape in the head of Taurus. You can't miss this cluster, because the V features the bright red giant Aldebaran (1st magnitude), which is Alpha Tauri (see Figure 12-2). Aldebaran actually is far beyond the Hyades cluster, but it appears in the same direction from Earth.

  The Hyades looks much bigger than the Pleiades because it's about 150 light-years from Earth versus about 400 light-years for the Pleiades.

- ✔ The Double Cluster, in Perseus, the Hero

  The Double Cluster is a beautiful sight through binoculars and especially through a small telescope. Its two clusters are NGC 869 and 884, each probably over 7,000 light-years from Earth. NGC stands for *New General Catalogue,* which was new when it first appeared in 1888 and didn't list any generals (or captains, colonels, and so on).

- ✔ The Beehive, in Cancer, the Crab

  The Beehive (Messier 44) is the main attraction of Cancer, a constellation composed of dim stars. It looks like a nice fuzzy patch with the naked eye and a swarm of many stars through binoculars. This cluster is about 500 light-years from Earth.

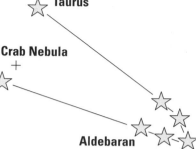

Figure 12-2:
Taurus
contains the
red giant
Aldebaran
(Alpha
Tauri).

For viewers in the Southern Hemisphere, the finest open clusters include

- NGC 6231, in Scorpius, the Scorpion

  NGC 6231 is a southern sky object, but you can readily see it from much of the United States in the evening during the summer. You need to be in a dark place with an unobstructed southern horizon. The observer Robert Burnham, Jr. described it as resembling "a handful of diamonds on black velvet."

- The Jewel Box, in Crux, the Cross

  The Jewel Box includes the bright star Kappa Crucis. Crux, popularly known as the Southern Cross, is a perennial favorite of viewers in the Southern Hemisphere. If you take a cruise through the South Seas, insist that the ship have an astronomy lecturer on board. (I will probably be available.) He or she can point out the Southern Cross; with binoculars, you can enjoy the fine sight of the Jewel Box.

A tight squeeze: Globular clusters

Globular clusters are the retirement homes of the Milky Way galaxy. They appear about as ancient as the galaxy itself (some experts think that the globular clusters were the first objects to form in the Milky Way), so they contain ancient stars, including many red giants and white dwarfs (see Chapter 11). The stars you can see in a globular cluster with your telescope are mostly red giants. With bigger telescopes, you can see orange and red main sequence dwarf stars. Only the Hubble Space Telescope and other very powerful instruments can pick out more than a handful of the much fainter white dwarfs in a globular cluster.

A typical globular star cluster contains a hundred thousand to a million or more stars, all packed in a ball (hence the term "globular") that measures just 60 to 100 light-years in diameter. The closer the stars are to the center, the more tightly they're packed in (see Figure 12-3). Its high degree of concentration and its great number of stars distinguish a globular cluster from an open cluster.

Figure 12-3:
Globular cluster G1 in the Andromeda Galaxy.

Courtesy of NASA

Another key difference is that open clusters are distributed across the galactic disk in a great flat pattern, but globular clusters are arranged spherically around the center of the Milky Way, with many high above and deep below the galactic plane. Most globular clusters are concentrated toward the galactic center, but many of the globulars you can easily see are well above or below the galactic plane.

The best globular clusters for viewing in the northern sky are

- Messier 13, in Hercules, representing the mythical character of the same name
- Messier 15, in Pegasus, the Winged Horse

You can spot both M13 and M15 with the naked eye under suitably dark sky conditions, but you may need to reassure yourself with binoculars or a small telescope, which show the clusters as fuzzy spots larger than stars. Use a star chart (such as *Norton's Star Atlas,* 20th Edition, by Ian Ridpath, PI Press, 2003) to locate the clusters.

Observers in the Northern Hemisphere are cheated out of the best globular star clusters, because by far the two biggest and brightest shine in the deep southern sky:

- Omega Centauri, in Centaurus, the Centaur
- 47 Tucanae, in Tucana, the Toucan

These clusters are spectacular sights through small binoculars and just about worth the trip to South America, South Africa, Australia, or other places where they can readily be seen, if you're an astronomer.

Be sure to check out the photo of a globular cluster in the color section.

Fun while it lasted: OB associations

OB associations are loose stellar groupings with dozens of stars of spectral types O and B (the hottest types of main sequence stars) and sometimes fainter, cooler stars (see Chapter 11 for more about spectral types). Unlike with open clusters and globular clusters, gravity doesn't hold OB associations together; over time, the stars move away from each other, dissolving the association like a limited partnership that has reached its limit. OB associations are located close to the galactic plane.

Many of the bright young stars in the constellation Orion are members of the Orion OB association. (See Chapter 3 for more about Orion.)

Taking a Shine to Nebulae

A nebula is a cloud of gas and dust in space ("dust" meaning microscopic solid particles, which may be made of silicate rock, carbon, ice, or various combinations of those substances; "gas" meaning hydrogen, helium, oxygen, nitrogen, and more, but mostly hydrogen). As I note in Chapter 11, some nebulae play an important role in star formation; others form from stars gasping on their deathbeds. Between the cradle and the grave, nebulae come in a number of varieties. (Check out a photo of a nebula in the color section of this book.)

Here are a few of the most familiar nebulae:

- **H II regions** are nebulae in which hydrogen is ionized, meaning that the hydrogen loses its electron. (A hydrogen atom has one proton and one electron.) The gas in an H II region is hot, ionized, and glowing, due to the effects of ultraviolet radiation from nearby O or B stars. All the large bright nebulae that you can see through binoculars are H II regions. (*H II* refers to the ionized state of the hydrogen in the nebula.)
- **Dark nebulae** are the dust bunnies of the Milky Way, consisting of clouds of gas and dust that don't shine. Their hydrogen is neutral, meaning that

the hydrogen atoms haven't lost their electrons. The term *H I region* refers to a nebula in which the hydrogen is neutral. It's another name for "dark nebula."

✔ **Reflection nebulae** are composed of dust and cool, neutral hydrogen. They shine by the reflected light of nearby stars. Without the nearby stars, they would be dark nebulae.

Sometimes a new reflection nebula appears suddenly, and you may discover it, as amateur astronomer Jay McNeil did. In January 2004, he found a new reflection nebula in the constellation Orion with a 3-inch refractor in his backyard, and professionals now call it McNeil's Nebula. But don't hold your breath; this type of discovery is very rare.

✔ **Giant molecular clouds** are the largest objects in the Milky Way, but they're cold and dark and scientists would've gazed right by them if not for the data gathered by radio telescopes, which can detect emissions of faint radio waves from molecules such as carbon monoxide (CO). Like all other nebulae, giant molecular clouds are mostly made of hydrogen, but scientists often study them by means of their trace gases, such as CO. The hydrogen in giant clouds is molecular, with the designation H_2, which means that each molecule consists of two neutral hydrogen atoms.

One of the most exciting nebular discoveries in recent decades showed that bright H II regions, such as the Orion Nebula, are just hot spots on the peripheries of giant molecular clouds. For centuries, people could see the Orion Nebula but had no idea that it's no more than a bright pimple on a huge invisible object, the Orion Molecular Cloud. But now we know. New stars are born in molecular clouds, and when they get hot enough, they ionize their immediate surroundings, turning them into H II regions. The part of a molecular cloud where the dust is thick enough to cut off the light of many or most of the stars behind the cloud, as visible from Earth, is called a dark nebula.

H II regions, dark nebulae, giant molecular clouds, and many of the reflection nebulae are located in or near the Milky Way's galactic disk.

Two other interesting types of nebulae are planetary nebulae and supernova remnants, which I cover briefly in the following sections (and in Chapter 11).

Picking out planetary nebulae

Planetary nebulae are the atmospheres of old stars that started out resembling the sun but then expelled their outer atmospheric layers, as the sun will do in the far future (see Chapter 10). The nebulae are ionized and made to glow by ultraviolet light from the hot little stars at their centers, which are all that remain of the former suns. Planetary nebulae expand into space and fade as they grow larger. They can be well off the galactic plane, unlike H II regions. (See a planetary nebula in the color section of this book.)

For decades, astronomers believed that many or most planetary nebulae were roughly spherical. But now astronomers know that most are *bipolar,* meaning that they consist of two round lobes projecting from opposite sides of the central star. The planetary nebulae that look spherical, such as the Ring Nebula in the constellation Lyra (see Figure 12-4), are bipolar, too, but the axis down the center of the lobes happens to point toward Earth (and so, like a dumbbell viewed end-on, they look circular). Astronomers took many years to figure this out, so maybe we were dumbbells, too.

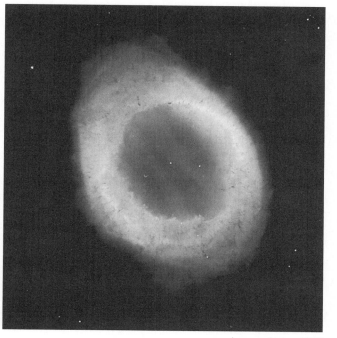

Figure 12-4:
The Ring Nebula in Lyra is bipolar but appears spherical from Earth.

Courtesy of NASA

Curious point: Respectively related and unrelated to planetary nebulae are *protoplanetary nebulae,* which are much studied by astrophysicists. One type of protoplanetary nebula is the early stage of a planetary nebula — a phase in the death of a star (not to be confused with the *Star Wars* Death Star). The other type of protoplanetary nebula is the birth cloud of a solar system's star and its planets. Yes, astronomers use the same term to refer to two completely different kinds of objects, but nobody's perfect. We may need another Edwin P. Hubble to bully us into some sensible nomenclature.

Correcting a galactic goof-up

Until the 1950s, astronomers used the term "nebula" to refer to a galaxy, because until the 1920s, they thought galaxies other than the Milky Way were nebulae in the Milky Way. Astronomers believed in the existence of only Earth's galaxy: the Milky Way.

It took a few dozen years for the change in understanding to prevail in the language of astronomy, so the authors of astronomy books have only recently stopped referring to the Andromeda Galaxy as the Andromeda Nebula.

Edwin P. Hubble, for whom the Hubble telescope is named, wrote the famous book *The Realm of the Nebulae*. He wrote all about galaxies, not nebulae as astronomers use the term now. Among his achievements, Hubble proved that the Andromeda Nebula is a galaxy full of stars, not a big cloud of gas. A former boxer, he fought in World War I, smoked a pipe, and supposedly bullied some of the other astronomers at Mount Wilson Observatory, but his discoveries are no bull.

Breezing through supernova remnants

Supernova remnants begin as material ejected from massive stellar explosions. A young supernova remnant is composed almost exclusively of the shattered remains of the exploded star that expelled it. But as the gas moves outward through interstellar space, it resembles a rolling stone that *does* gather moss. The expanding remnant creates a snowplow effect as it pushes along (see Chapter 11) and accumulates the thin gas of interstellar space. By the time it has aged — tens of thousands of years later — the remnant is overwhelmingly composed of this "plowed-up" interstellar gas, and the remains of the exploded star are mere traces. Supernova remnants expand along or near the galactic plane of the Milky Way.

Enjoying Earth's best nebular views

Nebulae are among the most beautiful sights through small telescopes. You need a good star chart, like those found in *Norton's Atlas,* and you should start with an easy target, like the Orion Nebula, which you glimpse by eye and binocular before homing in on it with your telescope. For H II regions like the Orion Nebula, a telescope with a low f/number, such as the ShortTube 80mm refractor from Orion Telescopes and Binoculars, may work the best (see Chapter 4, where I discuss using the scope for hunting comets, for more info on this particular tool). For smaller nebulae like the Ring Nebula, which I describe in the following list, the Meade ETX-90PE telescope (see Chapter 3) is a good choice for a beginner. The Meade has a computerized control system that points the telescope accurately at the little nebula (which you can't see with the naked eye).

The following are some of the best, brightest (or for dark nebulae, darkest), and most beautiful nebulae you can see from northern latitudes, including some southern-sky objects that aren't located very far south of the celestial equator:

✔ The Orion Nebula, Messier 42 (see Chapter 1), in Orion, the Hunter

An H II region, you can easily see the Orion Nebula with the naked eye as a fuzzy spot in the sword of Orion. It looks fine through binoculars and spectacular through a small telescope. The telescope also shows the Trapezium, a bright quadruple star (see Chapter 11) in the nebula.

✔ The Ring Nebula, Messier 57, in Lyra

The Ring Nebula is a planetary nebula high in the sky at northern temperate latitudes on a summer evening. Like all planetary nebulae, you need to use a star chart to find it with your telescope, unless you have a computer-assisted telescope such as the Meade ETX-90PE (see Chapter 3), which points right to the nebula at your command.

✔ The Dumbbell Nebula, Messier 27, in Vulpecula, the Little Fox

The Dumbbell Nebula, along with the Ring Nebula, is among the easiest planetary nebulae to spot with a small telescope. The best time for observation is in the summer and fall.

✔ The Crab Nebula, Messier 1, in Taurus

The Crab Nebula is the remains of a supernova that exploded in the year 1054, as seen from Earth. It appears as a fuzzy spot through a small telescope, but a big telescope shows two stars near its center. One star isn't associated with the Crab; it just appears along the same line of sight. The other star is the pulsar (see Chapter 11) that remains from the supernova explosion. It spins 30 times per second, and one or the other of its two lighthouse beams sweeps across Earth every $\frac{1}{60}$ of a second, with the same frequency as your 60-cycle, alternating-current household electricity. (I think that's a "powerful" analogy.)

✔ The North American Nebula, NGC 7000, in Cygnus, the Swan

The North American Nebula (the name comes from its shape) is a faint but large H II region that you can see with the naked eye on a summer evening at a very dark location with no moonlight. To glimpse it, use averted vision — look out of the corner of your eye.

✔ The Northern Coal Sack, in Cygnus, the Swan

The Northern Coal sSack is a dark nebula near Deneb, which is Alpha Cygni, the brightest star in Cygnus. You can identify it by eye as a dark blotch against the brighter background of the Milky Way.

You shouldn't skip the following nebulae located at moderate southern declinations; however, you can view them from many places in the Northern Hemisphere and of course from any place in the Southern Hemisphere:

✔ The Lagoon Nebula, Messier 8, in Sagittarius, the Archer

✔ The Trifid Nebula, Messier 20, in Sagittarius

The Lagoon Nebula and the Trifid Nebula are large, bright H II regions that you can see in the same field of view of your binoculars. Their best observation time is during summer evenings. A color photo shows that the Trifid has a bright red region and a separate, fainter blue region. The red area is the H II region and the blue zone is a reflection nebula.

Great nebulae of the deep Southern Hemisphere include

✔ The Tarantula Nebula, in Dorado, the Goldfish

The Tarantula Nebula is in the Large Magellanic Cloud galaxy, but it's such a huge and brilliant H II region that it's conspicuous to the naked eye for viewers in temperate southern and far southern latitudes. The Tarantula is another object to observe if you take a South Seas cruise, in addition to the Cross and the Jewel Box star cluster (see the section "Star Clusters: Galactic Associates" earlier in this chapter). Trust me, you won't carp at Dorado.

✔ The Carina Nebula, in Carina, the Ship's Keel

The Carina Nebula, near the huge, unstable star Eta Carinae (see Chapter 11), is a large, bright H II region.

✔ The Coal Sack, in Crux, the Cross

The Coal Sack, a dark nebula, is a large black patch, several degrees on a side, in the Milky Way. You can't miss it on a clear night with a dark sky, as long as you're deep in the Southern Hemisphere.

✔ The Eight-Burst Nebula, NGC 3132, in Vela, the Sail

The Eight-Burst Nebula is a planetary nebula visible in the far southern sky.

For some of the best color images of nebulae ever photographed, see the following Web pages of the Space Telescope Science Institute:

✔ The collection of nebular images originally distributed with press releases is at hubblesite.org/newscenter/newsdesk/archive/releases/image_category.

✔ A special selection of Hubble's best nebular images is at hubblesite.org/gallery/showcase/nebulae/index.shtml.

✔ The Hubble Heritage Image Gallery (with wonderful images of galaxies and other objects) is at heritage.stsci.edu/gallery/galindex.html.

Getting a Grip on Galaxies

A large galaxy consists of thousands of star clusters and billions to trillions of individual stars, all held together by gravity. The Milky Way fits this bill as a large spiral system. But galaxies come in many shapes and sizes (see Figure 12-5 for sketches of several main types).

The main types of galaxies, based on shape and size, are

 Spiral

 Barred spiral

 Lenticular

 Elliptical

 Irregular

 Dwarf

 Low surface brightness

I cover all these types in the following sections, as well as great galaxies to observe; the Milky Way's home, the Local Group; and even larger groups of galaxies, such as clusters and superclusters.

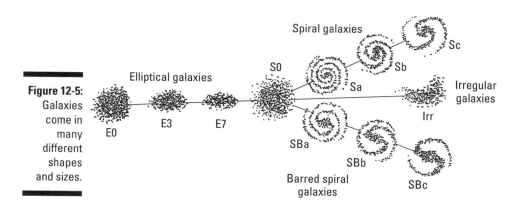

Figure 12-5: Galaxies come in many different shapes and sizes.

Surveying spiral, barred spiral, and lenticular galaxies

Spiral galaxies are disk-shaped, with spiral arms winding through their disks. They may resemble the Milky Way, or their arms may be wound more tightly than our galaxy's spiral arms (or less so). The central bulge of a star in another spiral galaxy may be more prominent or less prominent, compared to the spiral arms. The galaxies include the ones labeled by their *Hubble types,* Sa, Sb, and Sc, in Figure 12-5. (Yes, galaxy types are named after Edwin Hubble too.) As you go from Sa to Sc in that sequence (or beyond, to Sd), the galaxies' spiral arms are less tightly wound, and the central bulges are less prominent.

Spiral galaxies have plenty of interstellar gas, nebulae, OB associations, and open clusters, in addition to globular clusters. You can see a photo of spiral galaxies in the color section.

Barred spiral galaxies are spiral galaxies in which the spiral arms don't seem to emerge from the galaxy center; instead, they seem to protrude from the ends of a linear or football-shaped cloud of stars that straddles the center. This star cloud is called the *bar.* Gas from outer parts of the galaxy may be funneled toward the center through the bar in a process that forms new stars that make the galaxy's central bulge, well, bulge even more. These galaxies include the types labeled as SBa, SBb, and SBc in Figure 12-5. The sequence from SBa to SBc (and beyond to SBd) goes from galaxies with tightly wound arms and relatively large central bulges to those with open arms and small bulges.

Lenticular (lens-shaped) galaxies are flattened systems with galactic disks, just like spiral galaxies. They contain gas and dust, but they don't have spiral arms. These galaxies are marked with their type, S0, in Figure 12-5.

Examining elliptical galaxies

Elliptical galaxies are football-shaped, and that definition includes both the shapes of U.S. footballs and the shapes of soccer balls. Some ellipticals, in other words, are ellipsoidal in shape, roughly like a U.S. football, and some are spherical, like a soccer ball. Ellipticals can be beautiful sights, and I get a kick out of them. They contain many old stars and globular star clusters, but not much else. Elliptical galaxies are shown labeled with Hubble types from E0 to E7 in Figure 12-5. They form a sequence from the roundest ones at E0 to the flattest elliptical galaxies at E7.

Elliptical galaxies are systems in which star formation has largely or totally ceased. They have no H II regions, young star clusters, or OB associations. Imagine living in one of these dull galaxies, with nothing like the Orion Nebula to entertain you or give birth to new stars. And probably not much on television, either.

The production of new stars may have ended in an elliptical galaxy because all the interstellar gas was used up in making the stars already in the galaxy. Or star formation may have ended because something blew out or swept out all the remaining gas suitable for making more stars. I say "suitable" because some elliptical galaxies, although they show no H II regions or groups of young stars, do have some extremely hot gas — so thin and hot, in fact, that it shines only in X-rays. Gas in this state doesn't readily condense into a star.

A galaxy is a galaxy is a galaxy

Writing "galaxy" and "galaxies" over and over gets repetitive. But what's a good synonym for "galaxy?" Some uninformed folks (or their editors) write "star cluster" to vary their prose, but that's wrong. And a large group of galaxies isn't a "galactic cluster," which is a term meaning an open cluster of stars inside a galaxy. A large group is a *cluster of galaxies,* sometimes written as "galaxy cluster." The cluster is composed of galaxies, and is thus *galaxian,* but it isn't galactic.

And, to tell the truth, some elliptical galaxies do display a number of bluish star clusters, which appear to be very young globular star clusters, much younger than any in the Milky Way.

One leading theory of elliptical galaxies, or at least of *some* elliptical galaxies, is that they form through the collision and merger of smaller galaxies. The collision of two spiral galaxies, for example, could produce a large elliptical galaxy, and shock waves from the event may compress large molecular clouds in the spirals, giving birth to huge clusters of hot, young stars — perhaps the very bluish star clusters found in some ellipticals. But the collision of a small galaxy with a big spiral may only lead to the latter swallowing the former, making the spiral's central bulge even bigger.

As astronomers look out into space, we can see many examples of colliding and merging galaxies. The further back in time we look, the more prevalent the mergers seem to be. Apparently, galaxy collisions were common in the early universe and may have helped shape many of the galaxies that we see today.

Looking at irregular, dwarf, and low surface brightness galaxies

Irregular galaxies have shapes that tend to be, well, strictly irregular. You may find the glimmerings of a little spiral structure in one of them, or you may not. They generally have plenty of cool interstellar gas, with new stars forming all the time. And they usually appear smaller than full-sized spirals and ellipticals, with many fewer stars. You see an irregular galaxy with its type, Irr, in Figure 12-5.

Dwarf galaxies are just what the name implies: itty bitty galaxies that may be mere thousands of light-years across or less. The types of dwarf galaxies include dwarf ellipticals, dwarf spheroidals, dwarf irregulars, and apparently also (although their classification is still a bit controversial) dwarf spirals. Snow White only had seven dwarfs, but there may be billions of dwarf galaxies in the universe.

In our immediate neck of the woods, the Local Group of Galaxies, the most common galaxies are dwarf galaxies — just as in the Milky Way, the most common stars are the smallest stars, red dwarfs. The same probably holds true in the rest of the universe.

You don't see dwarf galaxies in Figure 12-5, because Edwin Hubble didn't include them when he made up the original diagram. He didn't include the next type that I list, the low surface brightness galaxies, because they hadn't been discovered yet. Hey, nobody's perfect.

Low surface brightness galaxies were recognized as a major variety in the 1990s. They can be as large as most other galaxies, but they barely shine at all. Although they have a full tank of gas, they haven't produced many stars, and therefore they don't appear as bright. Astronomers missed them for decades in surveys of the sky, but we're starting to pick them up now with advanced electronic cameras. Some very small low surface brightness galaxies have been found, which are the least luminous galaxies of them all; I call them "dim bulb galaxies." Who knows what else is out there that we haven't spotted yet?

Some astrophysicists think that much of the mass in the universe may be present in the form of low surface brightness galaxies that we haven't counted properly, like some groups of under-represented people in the U.S. Census.

Gawking at great galaxies

To enjoy telescopic views of galaxies, use telescopes like those I recommend in the section "Enjoying Earth's best nebular views" earlier in this chapter. Big galaxies like the Andromeda Galaxy or the Triangulum Galaxy are great sights through a low f/number telescope (see Chapter 3). For viewing smaller galaxies, I recommend a telescope with a computer control feature that points the telescope to exactly the right place in the sky. *Norton's Star Atlas* and other atlases show where the bright galaxies are among the constellations.

The best galaxies for viewing from the Northern Hemisphere include the following; when I mention the season of the year for best viewing, I mean the season in the Northern Hemisphere. (Remember, it's fall in the Northern Hemisphere when Brazilians are enjoying spring.)

✔ The Andromeda Galaxy (Messier 31; see Chapter 1), in Andromeda, a constellation named for an Ethiopian princess in Greek mythology

The Andromeda Galaxy is also called the Great Spiral Galaxy in Andromeda, and it was long known as the Great Spiral Nebula in Andromeda or just the Andromeda Nebula. It looks like a fuzzy patch to the naked eye. You can see it in the autumn evening sky. From a dark observing site, you can trace it across about three degrees — or about six times the width of the full Moon — on the sky with your binoculars.

Don't try to view the galaxy during the full Moon; wait until the moon is barely illuminated or is below the horizon. The darker the night, the more of the Andromeda Galaxy you can see.

✔ NGC 205 and Messier 32, in Andromeda

NGC 205 and Messier 32 are small, elliptical companion galaxies of the Andromeda Galaxy. Some experts call them both dwarf elliptical galaxies and some don't. (I wish they would make up their minds.) M32 is spheroidal in shape, and NGC 205 is ellipsoidal.

✔ The Triangulum, or Pinwheel Galaxy (Messier 33), in Triangulum, the Triangle

The Triangulum, or Pinwheel Galaxy, is another large, bright, nearby spiral galaxy, smaller and a little dimmer than the Andromeda Galaxy and also a fine sight through binoculars in the fall.

✔ The Whirlpool Galaxy (Messier 51), in Canes Venatici, the Venetian Hunting Dogs (see Figure 12-6)

The Whirlpool Galaxy is farther away and fainter than the Andromeda Galaxy and Triangulum Galaxy, but you get a more glorious view of it through a high-quality small telescope. The Whirlpool is a *face-on* spiral, meaning that the galactic disk is pretty much at right angles to our line of sight from Earth; we look right down (or up) on it. With the larger telescopes at a star party (see Chapter 2), you should be able to make out its spiral structure from a distance of about 15 million light-years. Messier 51 is where scientists discovered the spiral structure of galaxies. Look for it on a fine, dark evening in the spring.

✔ The Sombrero Galaxy (Messier 94), in Virgo, the Virgin

The Sombrero Galaxy is a bright, edge-on spiral galaxy. Its "brim" is the Sombrero's galactic disk. A dark stripe appears along the brim, because the band of dark nebulae or coal sacks in the Sombrero's galactic disk is edge-on to our line of sight. Look for The Sombrero in the spring, too; it may be almost three times farther away than the Whirlpool, but it still gives you a good view through a telescope.

The following list presents the finest galaxies for observers in the Southern Hemisphere:

✔ The Large and Small Magellanic Clouds (LMC and SMC) are irregular galaxies that orbit the Milky Way. The Large Cloud is not only larger but also closer to Earth. It orbits a mere 169,000 light-years (give or take) from us. In fact, scientists believed for many years that the LMC was the closest galaxy to the Milky Way. (Today, scientists know that a dim, miserable rendition of a galaxy, called the Sagittarius Dwarf Galaxy, is even closer. However, we can barely discern it in telescopic photos because the Milky Way is absorbing it. So long, Sagittarius, we hardly knew ye!)

Figure 12-6:
The Whirlpool Galaxy, photographed in ultraviolet light by the GALEX satellite.

Courtesy of NASA/JPL/Caltech

The LMC and SMC actually look like clouds in the night sky. They're that big and bright and are circumpolar in much of the Southern Hemisphere. In other words, at far southern latitudes, they never set below the horizon. If you go far enough south in South America or other places in the Southern Hemisphere, the LMC and SMC are visible on any clear night of the year. Sweep through them with binoculars and see how many star clusters and nebulae you can recognize.

✔ The Sculptor Galaxy (NGC 253) is a large, bright spiral galaxy and is one of the dustiest. Caroline Herschel, who also found eight comets, discovered it in 1783. Look for it with binoculars or a telescope on a dark autumn evening. You see it best from the Southern Hemisphere, but you can spot it from anywhere in the continental United States if you have a clear southern horizon.

✔ Centaurus A (NGC 5128) is a huge galaxy with a peculiar appearance: spheroidal, but with a thick band of dark dust across its middle. The galaxy is a powerful source of radio waves and X-rays and has been much studied with radio telescopes and with X-ray telescopes in orbiting satellites. Theorists have gone back and forth as to whether or not it's an example of colliding galaxies. I think that it has probably swallowed a smaller galaxy or two in its time, so watch from a safe distance. This object is most suitable for viewers in the Southern Hemisphere, with the best viewing in the autumn (spring in the Northern Hemisphere).

Discovering the Local Group of Galaxies

The Local Group of Galaxies, called Local Group for short, consists of two large spirals (the Milky Way and the Andromeda Galaxy), a smaller spiral (the

Triangulum Galaxy), their satellites (including the Large and Small Magellanic Clouds, as well as M32 and NGC 205), and about two dozen dwarf galaxies.

The Local Group isn't much as assemblages of galaxies go, but it's home and the largest structure that we on Earth are gravitationally bound to (meaning that Earth isn't flying away from the Local Group as the universe expands). Just as the solar system isn't getting any bigger — because the sun's gravity prevents the planets from moving outward or escaping — the Local Group holds strong because of the gravity of the three spiral galaxies and the smaller members. But all other groups and clusters of galaxies and distant individual galaxies throughout the universe outside the Local Group's gravitational pull *are* moving away from the Local Group at rates determined by a formula called *Hubble's Law* (named for the astronomer, not the telescope). Chapter 16 explains more about the movement away.

The Local Group is about one megaparsec across and centered near the Milky Way. A *parsec* is a dimension in space equal to 3.26 light-years, and *mega* means million, so the Local Group is about 3.26 million light-years or about 19 trillion miles wide. That dimension may sound large, but the distance is minuscule compared to the observable extent of the universe beyond.

Clusters and superclusters of galaxies are much larger than the Local Group, easily spotted across billions of light-years in space. But the majority of all the galaxies in the universe, at least the readily visible ones, are located in small groups with only dozens of members or less, like the Local Group (which has about 30). So we appear to be in an average condition, as galaxy neighborhoods go.

Checking out clusters of galaxies

Most galaxies may be in small groups like The Local Group, but as astronomers survey the distant heavens with professional observatory telescopes, the formations that stand out are the clusters of galaxies. Most prominent are the so-called *rich clusters,* with hundreds and even thousands of galaxy members, each with its own complement of billions of stars.

The nearest large cluster of galaxies is the Virgo Cluster, spread out across the constellation of the same name and adjacent constellations. The cluster is about 50 million light-years away and contains hundreds of known galaxies.

You can observe some of the biggest and brightest member galaxies of the Virgo Cluster with your own telescope. Messier 87 is one of the best sights: a spheroidal-shaped giant elliptical galaxy with a powerful jet of matter flying out at its center from the vicinity of a supermassive black hole. You can see M87 with amateur equipment, but not the jet at its center, unless you're a *very* advanced amateur. The galaxy appears to have swallowed some smaller ones, which may be why it's so big. Some galaxies like to start small and work their way up. Messier 49 and Messier 84 are two more Virgo Cluster giant ellipticals that you can observe, and Messier 100 is a large spiral galaxy in the cluster. Look for these galaxies on a dark evening in the spring in the Northern Hemisphere. Use a telescope with computer control that can home in on them for you. And if you don't trust computers, be sure you have a good star atlas that shows the galaxies.

Clusters of galaxies exist as far as our telescopes can see. At the limit of current technology in the early 21st century, we know of a few hundred billion galaxies in the observable universe, but nobody has counted them — at least nobody on our planet.

Sizing up superclusters, cosmic voids, and Great Walls

You may think that a large cluster of galaxies, up to 3 million light-years across, would be the max. But deep sky surveys indicate that most or all galaxy clusters are grouped into larger forms, called *superclusters*. The superclusters aren't held together by gravity, but they haven't fallen apart either. They appear to have long, filamentary shapes and flat, pancake-like shapes. A supercluster can contain a dozen clusters of galaxies, or hundreds of these clusters, and it can be 100 or 200 million light-years long.

We exist in the outer parts of the Local Supercluster, sometimes called the Virgo Supercluster, which is centered near the Virgo Cluster of Galaxies.

The superclusters seem to be positioned on the edges of huge, relatively empty regions of the universe called *cosmic voids*. The nearest one, the Bootes Void, is over 300 million light-years across. Many galaxies sit on its periphery, but we don't see very many inside it.

Astronomer Robert Kirshner discovered the Bootes Void. But when he was congratulated on the find, he reportedly said modestly, "It's nothing."

Some of the biggest superclusters, or groups of superclusters, are called *Great Walls*. The first Great Wall discovered is about 750 million light-years long. But other Great Walls, far out in the universe, may be larger. As far as astronomers know, the Great Walls don't display any Great Graffiti, but they have plenty to tell us about the origin of large structures in space and the early history of the universe. If only we understood the language.

Chapter 13

Digging into Black Holes and Quasars

In This Chapter

▶ Peering into the mysteries of black holes

▶ Getting the scoop on quasars

▶ Identifying different kinds of active galactic nuclei

Black holes and quasars are two of the most exciting and sometimes mystifying areas of modern astronomy, and lucky for us astronomers, the two subjects are related. In this chapter, I explain the connection between the two mysteries and provide you with information about active galactic nuclei, a group that quasars fall into.

You may never see a black hole through your own telescope, but I can guarantee that when you tell people you're an astronomer, they'll blurt out, "What's a black hole?" I mention black holes briefly in Chapter 11, but in this chapter, I offer you the "hole" treatment.

Black Holes: There Goes the Neighborhood

A *black hole* is an object in space whose gravity is so powerful that not even light itself can escape from within — which is why black holes are invisible.

You can fall in a black hole, but you can't fall out — you can't even get out if you want to (and you *would* want to). You can't even call home, so ET is lucky he lands on Earth, not in a black hole.

Anything that enters a black hole needs more oomph than it can ever have in order to get back out. The formal name for oomph is escape velocity. Rocket scientists use the term *escape velocity* to represent the speed at which a rocket or any other object must travel in order to escape Earth's gravity and pass into interplanetary space. Astronomers apply the term in a similar way to any object in the universe.

The escape velocity on Earth is 7 miles per second (11 kilometers per second). Objects with weaker gravity have slower escape velocities (the escape velocity on Mars is only 3 miles per second or 5 kilometers per second), and objects with more powerful gravity have higher escape velocities. On Jupiter, the escape velocity is 38 miles per second (61 kilometers per second). But the champion of the universe for escape velocity is a black hole. The gravity of a black hole is so strong that its escape velocity is greater than the speed of light (186,000 miles per second or 300,000 kilometers per second). Nothing, not even light, can escape from a black hole (because you need to travel faster than the speed of light to escape a black hole, and nothing — including light — travels faster).

Looking over the black-hole roster

Scientists can detect black holes when we see gas swirling around them that's too hot for normal conditions, when jets of high energy particles make their escape as though to avoid falling in, and when stars race around orbits at fantastic speeds, as though driven by the gravitational pull of an enormous unseen mass (which they are).

As I mention in Chapter 11, scientists recognize two main types of black holes:

- Stellar mass black holes have the mass of a large star (about three to a hundred times more massive than the sun) and they result from the deaths of such stars.

- Supermassive black holes, which are almost a million to a few *billion* times more massive than the sun, exist at the centers of galaxies and may have come from the merger of many closely packed stars when the galaxies formed. But nobody knows for sure.

Scientists discovered intermediate mass black holes, which have masses 500 to 1,000 times that of the sun, in 1999, but we don't know how they formed.

Poking around the black-hole interior

A black hole has three parts:

✔ The event horizon, which is the perimeter of the black hole

✔ The singularity, or the heart of the hole formed from the ultimate compression of all matter within it

✔ Matter that falls from the event horizon toward the singularity

The following sections describe these parts in more detail.

The event horizon

The event horizon is a spherical surface that defines the black hole (see Figure 13-1). After an object enters the event horizon, it can never get back out of the black hole or be visible again to anyone on the outside. And nothing on the outside can be seen from inside the horizon.

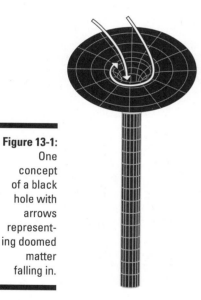

Figure 13-1: One concept of a black hole with arrows representing doomed matter falling in.

The size of the event horizon is proportional to the mass of the black hole. Make the black hole twice as massive, and you make its event horizon twice as wide. If scientists had a way to squeeze Earth down into a black hole (we don't, and if we did, I wouldn't tell how), our planet would have an event horizon less than ¾ of an inch (about 2 centimeters) across.

Table 13-1 offers a list of black hole sizes, in case you want to try some on.

Table 13-1		Black Hole Measurements	
Black Hole Mass in Solar Masses	**Diameter of Black Hole in Miles**	**Diameter of Black Hole in Kilometers**	**Comment**
3	11	18	Smallest stellar mass black hole
10	37	60	
100	370	600	Largest stellar mass black hole
1,000	3,700	6,000	Intermediate mass black hole
2.5 million	9.3 million	15 million	Supermassive black hole at Milky Way center
1 billion	3.7 billion	6 billion	Supermassive black hole in a quasar

As far as we know, no black holes are smaller than about three solar masses and 11 miles wide.

The singularity and falling objects

Anything that falls inside the event horizon moves down toward the singularity. It merges into the singularity, which scientists believe is infinitely dense. We don't know what laws of physics apply at these immense densities, so we can't describe what the conditions are like. We literally have a "black hole" in our knowledge.

Some mathematicians think that at the singularity there may be a *wormhole,* a passage from the black hole to another universe. The wormhole concept has inspired authors and movie directors to produce plenty of science fiction on the topic, but the writers and directors are just fishing. Most experts think that wormholes don't exist. And even if they do, scientists have no way to see them inside black holes or to worm our way down to them.

Another theory says that at the location where the hypothetical wormhole connects to another universe, there is a *white hole* — a place where enormous energy pours out into the other universe, like a gift from ours. This idea, too, seems wrong, but even if the theory *is* correct, we would have to journey to another universe (talk about frequent-flyer miles!) to see one.

Traveling to another universe (if one exists) is out of the question (at least for now). But the other possibility, of course, is to look for white holes in our universe, where a wormhole from another universe may emerge. Scientists have found no such thing.

Someone once suggested that quasars may be white holes. But today astronomers have perfectly good explanations for quasars (as I explain in the section "Quasars: Defying Definitions" later in this chapter), so as far as I'm concerned, astronomers are off the hook.

Surveying a black hole's surroundings

Here's what scientists observe in the vicinities of black holes:

1. **Gaseous matter falling toward the black hole swirls around in a flattened cloud called an *accretion disk*.**

2. **As the gas in the accretion disk gets closer to the black hole, it becomes denser and denser and hotter and hotter.**

   The gas heats up because the gravity of the black hole compresses it, a process that occurs because friction increases as the gas gets denser. (The process resembles the way air conditioners and refrigerators work: When gas expands, it cools, and when it gets compressed, it gets hotter.)

3. **As the hotter and denser gas approaches the black hole, it glows brightly; in other words, the accretion disk shines.**

   Radiation from the accretion disk can take many forms, but the most common type is X-rays. X-ray telescopes, such as the one on board NASA's large Earth-orbiting satellite, the Chandra X-ray Observatory, detect X-rays and allow scientists to pinpoint black holes. You can see the X-ray images from Chandra at chandra.harvard.edu, the Web site of the Chandra X-ray Center, by clicking on the photo album link.

So although you can't actually see a black hole through a telescope, you can detect radiation from the accretion disk of hot gas swirling around it — if you have an X-ray telescope that's up in space. X-rays don't penetrate Earth's atmosphere, so the telescope needs to be above the atmosphere.

Bare black holes may exist in space, with no gas swirling into them. If so, astronomers can't see them, unless they just happen to pass in front of a background star or galaxy under observation. In that case, you could infer that the black hole exists, because you see the effect of its gravity on the appearance of the background object. (You may see the background object get briefly brighter, for example, as I describe in Chapter 11 when I discuss microlensing.) But this situation would be a rare coincidence. Don't hold your breath waiting for a black hole to bare itself for your inspection.

Warping space and time

You can also describe a black hole as a place where the fabric of space and time are warped. A *straight line* — which is defined in physics as the path taken by light moving through a vacuum — becomes curved in the vicinity of a black hole. And, as an object approaches a black hole, time itself behaves oddly, at least as perceived by an observer at a safe distance.

Suppose that as you move at a safe distance, you launch a robotic probe toward a black hole from your spaceship. A big electric billboard on the side of the probe displays the time given by an onboard clock.

You watch the clock through a telescope as the robotic probe falls toward the black hole. What you see is that the clock runs slower and slower as the probe approaches the black hole. In fact, you never actually see the probe fall in. You see it get redder and redder as the light from the electric billboard is redshifted by the powerful gravity of the black hole — not due to the Doppler Effect (which I describe in Chapter 11), but due to a phenomenon called the *gravitational redshift*. The light of the billboard shifts toward longer wavelengths just as the Doppler Effect makes the light from a star moving away from the observer seem to shift toward longer wavelengths. After a while, gravity redshifts the glow of the electric billboard to infrared light, which your eyes can't detect.

Now consider what you'd see if you travel on board the falling probe. (Don't try this at home; in fact, don't try it anyplace.) You can watch the face of the clock from inside the probe, and you can peek back along your path through a window. You, the ill-fated onboard observer, see that the clock runs normally. You don't perceive that it runs slow at all. As you look out the window at the mother ship and the stars, everything seems to be blueshifted. You're blue at the thought that you can never go home again. You pass through an invisible boundary (the event horizon) around the black hole in no time at all. From then on, you can never see the outside again, nor can anyone outside see you.

The mother ship never sees you enter the black hole; you just appear to get closer and closer to the hole. But on the falling probe, you can tell that you've dropped right in. At least, you can if you're still alive. Ultimately, *tidal force*, an effect of the immense gravity, pulls anything that falls into a black hole apart; at least, along one dimension (the direction toward the singularity) you're pulled apart. To make matters worse, in the other two spatial dimensions, tidal force squeezes you together unmercifully.

If you enter the black hole feet first, tidal force stretches you out (if you're not already pulled apart) until you become tall enough to be drafted as a center in the NBA. But from bellybutton to back and from hip to hip, you get squeezed together like coal turning into diamond under immense pressure inside Earth, only worse. This experience is no gem.

Small or stellar mass black holes are the most deadly variety, just as some small spiders are more poisonous than big tarantulas. If you fall toward a stellar mass black hole, you're torn apart and squeezed together before you enter, and you never get to see the universe disappear before your end. But falling into a supermassive black hole is a happier experience. You get to fall inside the event horizon and see the universe black out before you suffer the tidal fate (or is it the fatal tide?).

Considering that black holes are all around us in the universe and that they have such fascinating and strange properties, you can see why scientists want to study them, but from a safe distance.

Quasars: Defying Definitions

Scientists have at least two definitions for quasars:

✔ **The original definition:** *Quasar* is an abbreviation or acronym for "quasi-stellar radio source" and means a celestial object that emits strong radio waves but looks like a star through an ordinary visible light telescope (see Figure 13-2). My friend, physicist Dr. Hong-yee Chiu, invented this term.

The original definition of quasar has become outdated because, at most, 10 percent of all objects that we now call quasars fit this definition. The other 90 percent don't emit strong radio waves. Astronomers call them radio-quiet quasars.

✔ **The current definition:** A *quasar* is a bright object at the center of a galaxy that produces about 10 trillion times as much energy per second as the sun and whose emissions are highly variable at all wavelengths.

Figure 13-2:
A quasar shines with 10 trillion times the power of the sun.

Courtesy of NASA

After decades of puzzling over quasars, astronomers have concluded that they're associated with giant black holes at the centers of galaxies. Matter that falls into the black hole releases enormous energy, and the observed energy sources are what astronomers call quasars.

Measuring the size of a quasar

All quasars produce strong X-rays; about 10 percent produce strong radio waves; and they all emit ultraviolet, visible, and infrared light. All the emissions can vary over years, months, weeks, and even over times as short as one day.

The fact that quasars often change significantly in brightness over the course of a single day indicates to scientists something of surpassing importance: A quasar must be no larger than about one *light-day,* or the distance that light travels through a vacuum over the length of a day. And a light-day is only 16 billion miles (26 billion kilometers) long, which means that a quasar, producing as much light as 10 trillion suns, or 100 times as much light as the Milky Way galaxy, isn't very much bigger than our solar system, which is a tiny part of our galaxy.

A quasar much bigger than a light-day couldn't fluctuate markedly in such a short time, any more than an elephant could flap its ears as rapidly as a hummingbird flutters its wings.

Getting up to speed on jets

Quasars that are strong radio sources often display *jets,* or long narrow beams in which energy shoots out of the quasars in the form of high-speed electrons and perhaps other rapidly moving matter. Often the jets are lumpy, with blobs of matter moving outward along the beams. Sometimes the blobs appear to move at more than the speed of light. This *superluminal motion* is an illusion related to the fact that the jets, in such cases, are pointing almost exactly at Earth; the matter in them actually moves close to light speed, but not faster than light.

 You can browse through the best images of jets from quasars as detected by radio telescopes in the Image Gallery of the National Radio Astronomy Observatory at `www.nrao.edu/imagegallery/php/level1.php`.

Exploring quasar spectra

Many books say that a quasar has very broad lines in its spectrum, corresponding to red and blue shifts of gas in turbulent motion within the quasar at up to 6,000 miles per second (10,000 kilometers per second). This statement isn't always true. Quasars come in a variety of types, and some don't have broad spectral lines. (See Chapter 11 for more on spectral lines.)

The broad spectral lines, however, are an important trait of many quasars and a clue to their relationship to other objects, as I describe in the next section.

Active Galactic Nuclei: Welcome to the Quasar Family

For years after the discovery of quasars, astronomers argued about whether they're located in galaxies or are separate from galaxies. Today, we know that quasars are indeed always located in galaxies, because technology has improved to the point that we can make a telescopic image that shows both a quasar and the galaxy around it. The latter is called the quasar's *host galaxy*. Because a quasar can be 100 times brighter than its host galaxy, or even brighter, hosts tend to get lost in the glare of their quasar guests, like the homeowner who puts up a celebrity for a night.

Electronic cameras, which can record a greater range of brightnesses in a single exposure than photographic film, made the discovery possible.

Quasars are an extreme form of what astronomers now call *active galactic nuclei (AGN)*. The term designates the central object of a galaxy when the object has quasarlike properties, such as a very bright starlike appearance, very broad spectral lines, and detectable brightness changes.

Sifting through different types of AGN

Scientists use the following main terms to describe active galactic nuclei (AGN):

- **Radio-loud quasars ("original quasars") and radio-quiet quasars (90 percent or more of quasars):** These two types are similar kinds of objects, with and without strong radio emission. They're located in spiral galaxies, like the Milky Way (see Chapter 12). We don't know if the quasars form in the galaxies or if the galaxies form around the quasars. No quasar is visible in the Milky Way galaxy, but we have detected a 2.5-million-solar-mass black hole at its center. That's a supermassive black hole, and I list it in Table 13-1.

- **Quasi-stellar objects (QSOs):** Some astronomers lump the radio-loud and radio-quiet quasars together as QSOs.

✔ **OVVs:** *Optically violently variable quasars* are quasars with jets that point directly at Earth. These quasars undergo even more pronounced rapid brightness changes than ordinary, garden-variety quasars. Think of firefighters struggling to direct a firehose at a person whose clothes are on fire. The water pressure may be unstable, with the water pulsing a bit. The stream from the hose may look pretty steady to spectators from the side, but the person on the receiving end feels every fluctuation in the flow as the oncoming water batters him. OVVs are the firehoses of quasardom — they make the biggest splash.

✔ **BL Lacs:** Astronomy lingo for *BL Lacertae objects,* BL Lacs as a group are AGN that resemble BL Lacertae. BL Lacertae changes in brightness, and for years scientists thought it was just another variable star in the constellation Lacerta (it looks like a star in photographs of the sky). They later identified it as a strong source of radio waves and eventually determined that BL Lacertae was the active nucleus of a host galaxy that had been lost in its glare until improved technology made it possible to photograph the galaxy.

Unlike most quasars, a BL Lac doesn't have broad spectral lines. And its radio waves are more highly polarized than those from ordinary radio-loud quasars (except the OVVs, which may be just extreme cases of BL Lacs) — *polarized* means that the waves have a tendency to vibrate in a preferred direction as they travel through space. Unpolarized waves vibrate equally in all directions as they move. At the ballpark, you can't tell the players without a scorecard, as the saying goes, and at the observatory, you have to check polarization to know your radio-loud quasars from your BL Lacs.

✔ **Blazars:** A term that covers both OVVs and BL Lacs. The two quasar types have many similarities. Both are highly variable in brightness, their jets point directly at Earth, and they're both radio-loud.

Do we really need a term to combine OVVs and BL Lacs? I'm not so sure. My friend Dr. Hong-Yee Chiu became famous among scientists for coining "quasar." His friend, Professor Edward Spiegel, coined "blazar" a few years later. If you discover a new kind of object, or write one of the leading studies on it, you may get to name it, too. Adding "ar" to your name isn't allowed; the term should be descriptive of the scientific properties of the object, not the astronomer.

✔ **Radio galaxies:** Galaxies with relatively dim active galactic nuclei that nevertheless produce strong radio emissions. Most of the strongest radio-emitting galaxies are giant elliptical galaxies. Often, they have beams or jets that transport energy from the AGN to huge lobes of radio emission, empty of stars, far outside and far larger than the host galaxy itself. There are usually two lobes on opposite sides of the galaxy.

✔ **Seyfert galaxies:** Spiral galaxies that have an AGN at their centers. A Seyfert AGN is like a quasar, with broad spectral lines and rapid brightness changes. It may be as bright as the host galaxy, but not 100 times brighter like a quasar, so the host isn't lost in the Seyfert nucleus's glare.

A Seyfert nucleus isn't a demanding guest; like a minor presidential candidate who visits a small town in Iowa without causing a big fuss. Local folks know that the candidate is around, but they tend to follow their daily routines instead of flocking downtown to greet the visitor. Carl Seyfert was the American astronomer who pioneered the study of these galaxies and their bright centers.

Examining the power behind AGN

All the different types of active galactic nuclei have one thing in common: They're powered by energy somehow generated in the vicinity of a supermassive black hole at their center.

Near the supermassive black hole, stars orbit the center of the host galaxy at immense speeds, which is how astronomers measure black holes' masses. With telescopes such as the Hubble, astronomers can determine the velocities of the orbiting stars or, sometimes, orbiting gas clouds by measuring Doppler shifts of the light from the stars or the gas (see Chapter 11 for more about the Doppler Effect). The speeds indicate the mass of the central object. Stars at a given distance from the center of a less-massive black hole orbit at a slower pace.

In a quasar or a radio galaxy of the giant elliptical type, the supermassive black hole often attains a billion or more solar masses. In Seyfert galaxies, the black hole mass is often around a million solar masses.

The black hole makes it possible for the AGN to shine, but only the matter falling into the black hole actually powers the glow. It may take matter with 10 times the mass of the sun falling into the black hole each year to make a quasar shine.

If no material falls into the black hole, it doesn't reveal itself by producing a bright glow, radio emission, high-speed jets, or strong X-rays. Like kids who depend on their school lunches for the energy to perform in class, the black holes only shine when matter falls into them at a sufficient rate. Supermassive black holes may be lurking at the centers of most galaxies, but in most cases, matter isn't feeding them, so astronomers only see quasars or other kinds of AGN in a small fraction of galaxies.

What came first: The black hole or the galaxy?

A recent discovery has brightened the day of every quasar fan. Experts have discovered a simple mathematical relationship between a supermassive black hole and the galaxy that surrounds it. The central region of most galaxies is called the *bulge*. Even a relatively flat spiral galaxy has a central bulge, which can be big, middling, or small, and an elliptical galaxy is considered all bulge. Astronomers have found that the mass of a black hole at the center of a bulge is always close to one-fifth of 1 percent of the mass of the whole bulge. It seems as though every galaxy has to pay a tax of 0.2 percent to its black hole. (I wish I got off that cheap from the IRS.)

This unexpected property of black holes and galaxies must be related to how they form, but astronomers aren't sure how. Does a big galaxy form around a big black hole? Or do big black holes form inside big, bulgy galaxies? Astronomers with inquiring minds are arguing over this now in a debate you may call the "battle of the bulges."

Proposing the Unified Model of AGN

The *Unified Model of Active Galactic Nuclei* is a theory that proposes that many kinds of AGN are in fact the same kind of object, which looks different when viewed from different angles. According to the Unified Model, when we look at AGN from different directions with respect to their accretion disks and their jets, they look different, just as a man you see face-on looks different than the same guy in profile. Everyone has a good side; from some angles, Jay Leno's chin doesn't seem so big. The theory also proposes that black holes are sucking in matter at different rates, so some AGN (which are getting more matter per second than others) are brighter than the others for that reason alone. Dozens of astronomers write papers on the Unified Model every year, some finding evidence for and some finding evidence against the theory.

I think that evidence points to real differences among the different types of AGN, but I also think that they have many basic similarities. Astronomers need more information before we can unite around the Unified Model or any other theory of AGN. In the meantime, what do you think? Your taxes pay for much of this research, which is underway in just about every developed nation, so you're entitled to an opinion.

Part IV

Pondering the Remarkable Universe

"Along with 'Antimatter' and 'Dark Matter', we've recently discovered the existence of 'Doesn't Matter', which seems to have no effect on the universe whatsoever."

In this part . . .

Read Part IV when you need a diversion, something to stir your mind with thought-provoking ideas and possibilities. Curl up with a cup of cider and read about the search for extraterrestrial intelligence (SETI). Have scientists found any evidence that little green beings are out there? Find out about dark matter, dark energy, and antimatter (yes, antimatter exists in the real world, not just in science fiction). And, when you feel ready, ponder the entire universe: how it began, its shape, and its future.

Chapter 14

Is Anybody Out There? SETI and Planets of Other Suns

In This Chapter

▶ Understanding Drake's Equation

▶ Exploring (and participating in) SETI projects

▶ Hunting for extrasolar planets

*T*he universe is both vast and varied. But do we share these starry realms with other thinking beings? Anyone who tunes in to *Star Trek* or frequents the local Cineplex already knows Hollywood's answer: The cosmos is cluttered with aliens (many of whom have managed to pick up a good amount of uninflected English).

But what do scientists say? Are there really extraterrestrial beings out there? Most researchers believe that the answer is yes. Some are even looking for evidence. Their quest is known as SETI (rhymes with "yeti"), the Search for Extraterrestrial Intelligence. (Other scientists have plans to search for traces of primitive life on Mars or some of the moons of the outer solar system, but SETI seeks advanced civilizations capable of broadcasting into space.)

Why are many scientists optimistic about the possible existence of aliens? Most of the upbeat attitude derives from the fact that our place in the cosmos is thoroughly unremarkable. The sun may be an important star to us, but it serves as a bit player in the universe. The Milky Way galaxy hosts 10 billion similar stars. If this number fails to impress you, note that over a hundred billion *other* galaxies are within range of our telescopes. The bottom line is that far more sun-like stars are sprinkled through the visible universe than we have blades of grass on Earth. To assume that our blade of grass is the only place where something interesting happens is (to put it gently) a bit gutsy. Distressing as it may be to our self-esteem, Earth may not be the intellectual nexus of the universe.

How can earthlings find our brainy brethren? You can't go visit their likely homes. Rocketing off to distant star systems, although a work-a-day staple of science fiction, is actually quite difficult. The speed of earthly rockets, an impressive 30,000 miles per hour, is less impressive when you reckon that it would take these craft a thousand centuries to reach Alpha Centauri, the nearest stellar stop on the tour of the universe. Faster rockets take less time, but they consume more energy — a lot more energy.

About 45 years have passed since astronomer Frank Drake made the first efforts to put us in touch with aliens. So far, our telescopes haven't snagged a single confirmed extraterrestrial peep. But keep in mind that until now the search has been limited. As technology (and, one hopes, funding) continues to improve, the chance for success increases. Someday soon astronomers may begin puzzling over a signal that comes from the cold depths of space. Maybe the signal will teach us interesting lessons, such as the meaning of life or at least all the laws of physics. But one thing is for sure: The signal will show us that we're not the only kids on the galactic block.

Using Drake's Equation to Discuss SETI

Although earthlings can't visit distant civilizations, astronomers are trying to find evidence of technically sophisticated space aliens by eavesdropping on their radio traffic. In 1960, astronomer Frank Drake attempted to listen in on cosmic communications by using an 85-foot (in diameter) radio telescope in West Virginia. If you've seen the movie *Contact,* you know that a radio telescope is similar to a seriously bulked up backyard satellite dish (see Figure 14-1). Drake connected his antenna to a new, sensitive receiver working at 1,420 MHz (located in what's called the microwave region of the radio spectrum) and then pointed the telescope at a couple of sun-like stars.

Drake didn't hear any aliens during his Project Ozma, but he provoked a great deal of enthusiasm within the scientific community. A year later, in 1961, the first major conference on SETI was held, and Drake tried to organize the meeting by distilling all the unknowns of the search into a single equation, now known as the *Drake Equation.* (For the mathematically inclined, I provide this simple little formula in the sidebar "Diving into the Drake Equation" later in this chapter.) Its logic is easy. The idea is to estimate N, the number of civilizations in our galaxy that use the radio airwaves now. N clearly depends on the number of suitable stars in the galaxy, multiplied by the fraction that have planets, multiplied by the number of . . . well, you can read about it in the sidebar.

Drake's Equation is truly seductive, and you may want to impress strangers by rattling it off at a dinner party. But although scientists may know or can safely guess at the values of the first few terms in the equation (such as the rate at which stars capable of hosting planets form and the fraction of such stars that actually have planets), we don't have any real knowledge of such details as the fraction of life-bearing planets that can develop intelligent life or the lifetime of technological societies. So Drake's Equation still doesn't have an "answer." But it is a great way to organize the SETI discussion.

Figure 14-1:
A radio telescope is no more than a specialized antenna. With the right kind of receiver, astronomers can listen for signals from other societies.

SETI Projects: Listening for E.T.

Most of the modern SETI (Search for Extraterrestrial Intelligence) efforts follow in Frank Drake's footsteps. In other words, they use large radio telescopes in an attempt to eavesdrop on signals from alien civilizations.

Why use radio? Radio waves move at the speed of light and easily punch through the clouds of gas and dust that fill the space between the stars. In addition, radio receivers can be quite sensitive. The amount of energy required to send a detectable signal from star to star (assuming that aliens wield a transmitting antenna at least a few hundred feet in size) is no more than what your local television station pumps out.

Assuming that researchers do get an interstellar ping, how do they recognize it? They don't look to receive the value of *pi* or some other simple message that proves the aliens have completed junior high. SETI researchers simply look for narrow-band signals.

Narrow-band signals occur at one narrow spot on the radio dial. Only a transmitter can make narrow-band emissions. Quasars, pulsars, and even cold hydrogen gas all make radio waves, but their natural static is spread out in

Diving into the Drake Equation

Scientists often use Frank Drake's nifty little formula as the basis for discussions about SETI and the chances that humans will ever make contact with extraterrestrial intelligent life. The equation is quite simple and doesn't require any math beyond what you mastered in the eighth grade.

The equation computes *N*, the number of broadcasting civilizations active in the Milky Way galaxy. As with the Bible, several versions of Drake's Equation exist, but here's the usual formulation, in all its gory glory:

$$N = R*f_p n_e f_i f_i f_c L$$

- *R** is the rate at which long-lived stars, suitable for hosting habitable planets, form in the galaxy. Because the Milky Way has roughly 300 billion stars and is approximately 13 billion years old, this number is about two per year. (About one in ten stars is close enough in size and brightness to the sun to merit orbiting, habitable planets.)

- f_p is the fraction of good stars — stars that can have habitable planets — that actually *have* planets. Astronomers don't know what this number is, but we know it's at least 10 percent and possibly much higher.

- n_e is the number of planets per solar system that can incubate life. In our solar system, the number is at least one (Earth) and could

be more if you count Mars and some of the moons of Jupiter and Saturn. But in another system, who knows? A typical guess is one.

- f_i is the fraction of habitable planets that actually develop life. We can reasonably assume that many of them do.

- f_i is the fraction of planets with life that develop intelligent life. This number is controversial, of course, because intelligence may be a rare accident in biological evolution.

- f_c gives the fraction of intelligent societies that invent technology (in particular, radio transmitters or lasers). Probably most of them do.

- *L*, the final term, is the lifetime of societies that use technology. This term is a matter of sociology, not astronomy, of course, so your guess is as good as the author's. Maybe better.

The number *N* depends on your choice of values for the various terms. Pessimists think that *N* may be only one (we're alone in the Milky Way galaxy). Carl Sagan figured it to be a few million. And what does Drake say himself? "About 10,000." Moderation in all things. You can visit the site www.seti.org to tinker with the Drake Equation by entering your own values and calculating *N*.

frequency — splattered all over the radio spectrum. Narrow-band signals are the mark of transmitters. And transmitters are the mark of intelligence. It takes intelligence (not to mention a soldering iron) to build a transmitter.

Another criterion that SETI researchers insist upon before they can claim a true, alien broadcast is that the signal be persistent. In other words, every time they point their telescopes at the source of the signal, they find it. If their meters register only once, the signal is impossible to confirm. They may count it as interference from telecommunication satellites, a software bug, or an ambitious college prank.

In the following sections, I discuss several SETI projects and explain how you can help with the search.

The flight of Project Phoenix

The most sensitive SETI experiment researchers have conducted so far is Project Phoenix, run by the SETI Institute in Mountain View, California, from 1995 to 2004. This project was a successor to a NASA SETI program Congress halted in 1993 (indeed, ever since that date, all SETI efforts in the United States have been privately funded).

Project Phoenix trained its sights on individual stars, in what is known in the SETI trade as a *targeted search*. Other projects use telescopes to sweep large tracts of the sky. Of course, those broad sweeps let scientists examine more of the heavens, but by concentrating only on nearby, sun-like stars, a targeted search achieves much greater sensitivity. In other words, it can find far weaker radio signals. Researchers carried out Project Phoenix on a handful of different telescopes, including the 1,000-foot (in diameter) Arecibo radio telescope (in Puerto Rico), the mother of all antennas (see Figure 14-2).

Figure 14-2:
A view of the massive Arecibo radio telescope in Puerto Rico that participated in Project Phoenix.

Courtesy of Seth Shostak

Phoenix (and many other SETI experiments) looked for signals in the microwave region of the radio dial. Microwaves, aside from their ability to make leftovers palatable, are the preferred "hailing channel" of the SETI crowd for two reasons:

 ✔ The universe is rather quiet at microwave frequencies — you encounter less natural static, a fact that E.T. also knows.

 ✔ A natural signal generated by hydrogen gas occurs at 1,420 MHz, a frequency located in the microwave region. Because hydrogen is far and away the most abundant element in the cosmos, every alien radio astronomer should be aware of this natural marker — and may be tempted to get our attention (or the attention of any other civilization in space) by sending out a signal near to its frequency on the dial.

But facing facts, scientists really don't know *exactly* where the extraterrestrials may tune their transmitters. To cover as much of the dial as possible, Project Phoenix checked out many millions of channels at once (and over the course of time, billions of channels for each targeted star).

When Project Phoenix finally stopped its observing program in the spring of 2004, it had carefully examined about 750 sun-like star systems. No persistent, clearly extraterrestrial signal was found. But this effort taught researchers how to build an instrument that could, in a few decades, check out a million star systems or more. Those lessons have resulted in the efforts to construct the Allen Telescope Array (see the next section).

Space scanning with other SETI projects

Today, several SETI programs dot the astronomy landscape:

 ✔ The *Search for Extraterrestrial Radio Emissions from Nearby Developed Intelligent Populations* (nicknamed *SERENDIP*), conducted from the University of California, Berkeley, uses the Arecibo telescope in a "piggyback" mode, collecting data from whatever direction the telescope points. In this way, researchers can use the telescope for SETI even when other astronomers are studying pulsars, quasars, or other natural objects. This apparently aimless approach pays off handsomely in observing time: SERENDIP collects data almost every day, all day.

 ✔ *Southern SERENDIP* is run by the SETI Australia Centre in New South Wales. The centre uses a 210-foot radio telescope at Parkes, in the sheep and mosquito country a few hundred miles west of Sydney. The southern version is also a piggyback experiment, in that astronomers other than the SETI researchers control the aim of the antenna.

✔ The SETI Institute and the University of California at Berkeley are building a new radio telescope, called the Allen Telescope Array, designed
from the get-go for efficient SETI searching. The scope will consist of 350
small (6 feet in diameter) antennas spread over a half-mile of California
real estate (see Figure 14-3). Its first SETI experiment will be to scan the
densest parts of the Milky Way galaxy for telltale signals.

The Allen Telescope Array's 350 antennas are like synchronized swim-
mers: They all point in the same direction. But unlike many past SETI
instruments, the antennas can observe several candidate star systems
simultaneously. This, of course, speeds up the search, as does the fact
that the array can function 24/7 for SETI. This project is, without doubt,
the most ambitious SETI telescope project to date, and it should be completed sometime before 2010.

Figure 14-3:
Upon
completion,
the Allen
Telescope
Array will
consist
of 350
antennas
spread
across a
half-mile in
California.

Courtesy of the SETI Institute

In addition to these SETI efforts, a growing number of so-called optical SETI
experiments are showing up. Instead of hunting for radio broadcasts, optical
projects look for very brief but intense flashes of laser light that may come
our way from a society keen to get in touch. Optical SETI relies on conventional mirror-and-lens telescopes, outfitted with high-speed electronics to

sense and record any alien light bursts. Optical SETI experiments are underway at the University of California at Berkeley, the Lick Observatory (a University of California observatory), and at Harvard University.

Although a light signal from a distant world may be obscured by the light from that planet's sun, it's pretty easy to focus the laser light with a mirror. Such an optical transmitter could outshine a star — for the billionth of a second that the flash is on! So it makes sense to look for messages sent in this flashy way, and optical SETI researchers have already trained their telescopes on a few thousand nearby stars. So far, no dice.

You can find links to the Web sites of all major SETI programs at the SETI Institute's site (www.seti.org) or at the Planetary Society's Web page (seti.planetary.org).

SETI programs want you!

You too can join in the following SETI projects:

- SETI@home is a part of project *SERENDIP* (see the previous section). If you go to its site (setiathome.ssl.Berkeley.edu), you can download a snazzy screensaver free of charge. After you install the software on your computer, your modem connects to a server at Berkeley to retrieve a chunk of SETI data. The screensaver software crunches away on the data, looking for signals. After a few days (depending on how often you leave your computer to its own devices), the results upload to the server.

- The SETI League, headquartered in scenic New Jersey, recruits radio amateurs to use their backyard dishes in the search for cogitating aliens. With technical help from the League, you can become a do-it-yourself SETI searcher, building the necessary receivers and downloading the software required to hunt for signals. If you're an electronics propeller-head, this could be for you. Visit www.setileague.org.

Although your chances of finding a telltale tone are small, the chance still exists. Who knows? You may enjoy sharing a plate of meatballs with the Swedish king after collecting your Nobel Prize. Or maybe you don't like to share.

Finding Extrasolar Planets

One term of Drake's famous formulation is *fp*, the fraction of sun-like stars that sport planets (see the sidebar "Diving into the Drake Equation," earlier in this chapter, for more details). Astronomers have believed for decades that planets are plentiful, simply because the birth of a star is inevitably accompanied by leftover material — a messy residue of gas and dust that can turn into small, orbiting worlds.

But to actually find planets around stars is tough. You can't simply point a telescope in the direction of a nearby star and hope to see its planets. The orbiting bodies are too dim and too close to a blinding light source (their sun). To grasp the full existential challenge of the problem, imagine trying to see a marble located 30 yards from a light bulb at a distance of 10,000 miles.

Despite these daunting difficulties, astronomers *have* found *extrasolar planets* (planets outside of our solar system that orbit stars other than our sun) — not by picking them out in photos, but by

- ✔ Measuring the slight dimming caused when they pass in front of their sun
- ✔ Carefully monitoring the slight dance their sun makes because of their presence

The first technique (dimming) takes advantage of the fact that if another solar system is by chance oriented the right way, the planets will — once per orbit — cross in front of their home star as seen from Earth. The mini-eclipses obviously reduce the brightness of the star, if only for a few hours. The dimming is slight: only about 1 percent, even for a hefty Jupiter-sized world. An attentive astronomer with good equipment, however, can notice the difference.

Astronomers have found a few planets with this so-called *transit* technique, but most of the alien worlds discovered since 1995 have been uncovered by the second technique: measuring the small motions of the host stars.

Planets and stars orbit their common center of mass, and this arrangement means that both objects move. As they orbit around under the influence of their mutual gravitational attraction, the star pulls on the planet, making the planet move, and the planet pulls on the star, making the star move. The planet has much less mass than the star, so the so-called reflex motion of the star usually isn't much — perhaps only 50 miles an hour (compared with the planetary motion, which may be 10,000 miles an hour or more). But by using sensitive *spectroscopes* — devices that break up the incoming light into its various colors, like a prism — mounted on large telescopes, astronomers can hunt for the small Doppler Effect (see Chapter 11) that the slow stellar wobble produces in the star's light. And they've already managed to find more than 100 stars whose lazy dancing betrays orbiting planets.

The following sections cover several interesting extrasolar planets and plans for continuing the search.

51 Pegasi's hot partner

The credit for discovering the first extrasolar planet belongs to two Swiss astronomers, Michel Mayor and Didier Queloz, who announced their find in the fall of 1995. The discovery caused a great deal of consternation in the research community, mainly because the new planet whips about its star (51 Pegasi) at a breakneck pace. A complete orbit takes only about four days. Consequently, astronomers deduced that the planet orbits a trifling 5 million miles from its host star (see Figure 14-4), or about eight times closer than Mercury is to the sun. The findings also imply that the temperature of the world is roughly 1,000°C. The size of 51 Pegasi's stellar wobble indicates that the planet's mass is at least half that of Jupiter. For obvious reasons, astronomers soon dubbed the new planet a *hot Jupiter*.

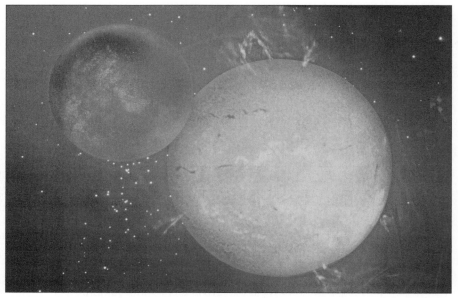

Figure 14-4:
An artist's
concept of a
hot Jupiter
orbiting
the star
51 Pegasi.

Courtesy of Seth Shostak

In the decade following the discovery of 51 Pegasi's warm little world, astronomers have found new planets at the rate of more than one a month, nearly all with spectroscopic measurement of Doppler shifts. Quite a few of the new worlds are also hot Jupiters — massive planets that hug their home stars tighter than a loving mom.

But astronomers don't believe that any of these hot and heavy worlds were born in their present, toasty orbits. Large planets have a far easier time forming in the dim suburbs of a solar system. The colder temperatures and endless expanse of material in these nether regions encourage the rapid conglomeration of icy debris into large worlds. But upon birth, the planets' interactions with the leftover debris may cause them to wander from home and gravitate inward to the hellish domains of their scorching home stars.

Still, most of the newly discovered planets don't orbit their stars as tightly as the planet of 51 Pegasi. This is good news for anyone who likes to think that there may be other solar systems in the universe like our own. Many of the first extrasolar planets scientists found are hot Jupiters, which caused some people to worry that our solar system (where the big planets like Jupiter are far from the sun) is freakish and rare. But as the search continues, astronomers are finding more and more extrasolar planets that have orbits similar to the planets of our solar system.

No one is quite sure what prevents hot Jupiters from going "all the way" and careening into their host stars. One possibility is that these planetary heavyweights raise waves of hot gas on the star's outer surface, and the gravitational effects of these tides halt the planet's inward spiral. But that's still a theory, and astronomers candidly admit that both the birth and ultimate fate of hot Jupiters are phenomena that we don't yet understand.

The Upsilon Andromedae system

In 1999, Geoff Marcy, Paul Butler, and other collaborators (who have discovered many of the new planets detected since 1995) added to the planet-finding excitement by claiming that not one, but three large planets are in orbit around the star Upsilon Andromedae. The group made this discovery through careful analysis of the star's subtle wobbling motions.

Upsilon Andromedae, an F-type star 44 light-years from Earth, thus became the first *normal star* (a shining, nuclear furnace type of star) other than the sun known to have a genuine solar system. (See Chapter 11 for more about different types of stars.) The planets themselves are hefty, weighing in at greater than 0.7, 2.1, and 4.6 times the mass of Jupiter. They don't all hug the star, however: The outer two planets have orbits comparable in radius to those of Venus and Mars.

Continuing the search for planets suitable for life

Although it reassures people who search for extraterrestrials to know that E.T. has plenty of homes to phone, the new planet discoveries are also a bit disconcerting. After all, hot Jupiters (or, for that matter, cold Jupiters) aren't likely places for biology to cook up, because water on such worlds would either boil or freeze, and liquid water is what we think that all life — including alien life — requires. If these oversized planets are typical of the galaxy's complement of worlds, we shouldn't expect much cosmic company.

But such a scenario is unlikely. The technique that scientists use to find most of the new planets — looking for wobbles by taking advantage of the Doppler Effect in the light from stars — is best for uncovering giant worlds that orbit close to their home stars. You can compare the search for planets so far to a reconnaissance of the African savannas from a helicopter. You can see the elephants and rhinos, but you miss the mice and mosquitoes. Scientists find big planets because we *can* find big planets. Small worlds are probably plentiful, but until researchers build some new types of telescopes, discovering the small guys is difficult.

In 2007, NASA plans to launch the Kepler mission: a space-based telescope whose task is to find out if small planets really are as common as phone poles. It will stare for four years at a patch of sky containing 100,000 relatively nearby stars, hoping to catch sight of the periodic dimming caused by encircling worlds. The expectation is that Kepler will find many dozen Earth-size worlds with the transit technique, but of course we have to wait to see what really happens.

If planets the size of Earth are abundant, the next step is to find out if any of them support life. The answer may be within reach if proposed new space telescopes are built — instruments such as the Terrestrial Planet Finder (NASA) or Darwin (European Space Agency). These high-flying, high-tech scopes, which researchers hope to launch by 2010 or 2012, would be able to actually catch some of the light from extrasolar planets and, with a bit of simple spectral analysis, determine what compounds make up their atmospheres. If scientists find a lot of oxygen or methane in the air of some far-off world, we may have good reason to suspect the presence of life. Needless to say, building space-based telescopes is much easier than sending a Federation Starship out on a reconnaissance mission.

If you want to know the latest and greatest in the search for extrasolar planets, you can find the facts at `cfa-www.harvard.edu/planets`, which also has links to many other related sites.

Dr. Seth Shostak, Senior Astronomer at the SETI Institute in Mountain View, California, contributed this chapter.

Chapter 15

Delving into Dark Matter and Antimatter

* *

In This Chapter

▶ Discovering the concept of dark matter

▶ Searching for evidence of dark matter

▶ Becoming attracted to antimatter

* *

Stars and galaxies set the night sky aglow, but these glittering jewels account for only a tiny portion of the matter in the cosmos. There's more to the universe than meets the eye — much more.

This chapter introduces you to the concept of dark matter, tells you why astronomers are convinced that the stuff must exist, and describes experiments that may shed light on the nature of this mysterious, invisible material. I also discuss another exotic type of matter in the universe: antimatter. Yes, antimatter exists outside of science fiction, but the real-world version is every bit as fascinating as the sci-fi books, television shows, and movies suggest.

Dark Matter: Understanding the Universal Glue

As far back as the 1930s, astronomers found hints that at least 90 percent of the mass in the universe doesn't emit, reflect, or absorb light.

The invisible material, known as *dark matter*, serves as the gravitational glue that keeps a rapidly rotating galaxy from flying apart and enables fast-moving galaxies in a cluster to stick together. Dark matter also seems to have played a crucial role in the development of the universe as we know it today — a spidery network of immensely long superclusters of galaxies separated by giant voids (see Chapter 12). Indeed, dark matter may determine the ultimate fate of the cosmos.

If the arguments in the following sections are correct, at least 90 percent — perhaps even 99 percent — of the matter in the universe is dark. What a humbling thought. The universe you see when you peer through a telescope or when you look up at the night sky teeming with stars and galaxies is just a tiny fraction of what's out there. To borrow a nautical analogy, if galaxies are like sea foam, dark matter is the vast, unseen ocean on which it floats.

Gathering the evidence for dark matter

The first hint that the universe contains dark matter appeared in 1933. While examining the motions of galaxies within a large cluster of galaxies in the constellation Coma Berenices, astronomer Fritz Zwicky of the California Institute of Technology found that some galaxies move at an unusually high speed. In fact, the galaxies of the Coma Cluster move so rapidly that all the visible stars and gas in the cluster can't possibly keep the galaxies gravitationally bound to one another, according to our known laws of physics. Yet somehow the cluster remains intact.

Zwicky concluded that some sort of unseen matter must exist within the Coma Cluster to provide the missing gravitational attraction.

As surprising as this conclusion was, dark matter didn't make headlines for several more decades. Many astronomers figured that after they studied the motions of galaxies in greater detail, the rationale for the existence of the invisible material would disappear. Instead, in the 1970s, evidence for dark matter became more compelling. Not only do clusters of galaxies seem to contain the stuff, but also individual galaxies. The following sections describe the main arguments in favor of dark matter.

Dark matter makes stars orbit oddly

Vera Rubin and Kent Ford of the Carnegie Institution of Washington, D.C. were studying the motions of stars in hundreds of spiral galaxies when they obtained a result that seemed to fly in the face of conventional physics. A spiral galaxy resembles a flattened fried egg, with most of its mass seemingly concentrated in the yolk — astronomers call this the *bulge* (as I explain in Chapter 12). Images reveal that the visible mass of a spiral diminishes rapidly with increasing distance from the bulge.

Scientists naturally expect the stars in a spiral galaxy to orbit their massive center in the same way that planets in our solar system orbit the sun. Obeying Newton's law of gravity, the outer planets, such as Pluto and Neptune, orbit the sun more slowly than the inner planets, such as Mercury, Venus, and Earth. Therefore, stars in the outskirts of a spiral galaxy should orbit more slowly than stars nearer the center. But that's not the result Rubin and Ford arrived at.

In galaxy after galaxy, their observations revealed that the outlying stars orbit rapidly, just like the inner stars. With so little visible material in the outer regions, how do the outlying stars manage to zip around so fast and still stay bound to the galaxy? They should escape from their galaxy, given their speed.

The astronomers concluded that *visible matter* — the stars and luminous gas that show up on telescopic photographs — makes up only a small portion of the total mass of a spiral galaxy.

Although the visible mass is indeed concentrated at the center, a vast quantity of other material must extend far beyond. Each spiral galaxy must be surrounded by an immense *halo* of dark matter. And to exert enough of a gravitational tug on the stars in the visible outskirts of the galaxies, the dark matter must exceed the visible matter by at least a factor of 100 in mass. Other types of galaxies (elliptical and irregular) also have dark matter halos.

Cold dark matter makes the universe take its lumps

Cosmologists (scientists who study the large-scale structure of the universe and its formation) also point to dark matter to explain a fundamental puzzle about the universe: How did it evolve from a nearly uniform soup of elementary particles in the aftermath of the Big Bang (see Chapter 16) to reach its present lumpy structure of galaxy clusters and superclusters?

Even though 13.7 billion years have passed since the birth of the universe, scientists don't believe that enough time has passed for visible matter to coalesce on its own into the huge cosmic structures we see today.

To solve this cosmological conundrum, some experts state that the universe contains a special type of dark matter, called *cold dark matter,* that moves more slowly and gathers into clumps more quickly than ordinary, visible matter. Responding to the tug of this exotic material, the ordinary matter formed stars and galaxies within the densest concentrations of this dark matter. This theory explains why every visible galaxy seems to be embedded in its own dark-matter halo.

Is the cold dark matter theory correct? It appears to be in broad agreement with the facts about the universe, as far as scientists know them, but the agreement isn't perfect. For example, the theory predicts that hundreds of tiny satellite galaxies will surround a big galaxy like the Milky Way. But we don't see all that many satellite galaxies. The predictions of the theory may need work, or we may need a better theory of dark matter. Or maybe there are small, faint galaxies all around us that we haven't found yet. They may be the "dim bulbs" of the universe.

Dark matter is critical to the universal density

Astronomers believe in dark matter for yet another cosmic reason: The universe, on large scales, looks the same in all directions and has an overall smoothness. This consistency in appearance and smoothness indicates that the universe has just the right density of matter, called the *critical density* (which I explain in Chapter 16). The total amount of visible matter that we observe in the universe isn't nearly enough to achieve critical density. Dark matter takes up the slack.

Debating the makeup of dark matter

Okay, astronomers have plenty of good reasons to believe in dark matter. But what the heck is the stuff anyway?

Broadly speaking, astronomers divide the possible kinds of dark matter into two classes: baryonic dark matter and oddball dark matter.

Baryonic dark matter: Lumps in space

Some dark matter may consist of the same stuff that the sun, planets, and people are made of. This kind of dark matter would be part of the family of *baryons,* a class of elementary particles that includes the protons and neutrons found in the nuclei of atoms.

Baryonic dark matter includes all hard-to-see material that's made of known kinds of matter, including asteroids, brown dwarfs, and white dwarfs (I describe the dwarfs in Chapter 11). Yes, scientists can detect asteroids in our solar system and white dwarfs and brown dwarfs nearby in the Milky Way. But far out in the galactic halo, these objects may be undetectable with present equipment. Such hypothetical objects in the galactic halo, referred to as *MACHOs (massive compact halo objects),* may account for the dark matter halos surrounding individual galaxies. (I cover the search for MACHOs later in this chapter.) But we don't see nearly enough of them to account for the development of large-scale structure in the cosmos. I think this theory is probably wrong, and the scientists who proposed it have to take their lumps.

Oddball dark matter: Stranger still

Alternatively, dark matter may consist of an abundance of exotic subatomic particles that bear little or no resemblance to baryons. These particles include *neutrinos,* which do exist (see Chapter 10 for more about them), and others with names such as *axions, squarks,* and *photinos* that physicists have dreamt up without proof of their existence. (Yes, experiments are underway, but no one has captured an axion or any other hypothetical dark matter particle thus far.)

During the *Big Bang* at the birth of the universe (see Chapter 16), a zoo of weird, dark-matter particles may have formed and a few may have survived. The zoo may include the axion, a kind of miniature black hole 100 billion times lighter than an electron. Even though axions are featherweights, if enough exist, they could contribute significantly to the cosmic mass. Recent experiments suggest that the neutrino (a particle that scientists thought to have, perhaps, zero mass) does have a very small but real mass and may account for a small portion of the dark matter.

Other candidates for oddball dark matter are heavier — about 10 times the mass of the proton — but still insubstantial in terms of making up the dark matter in the universe, unless they occur in large numbers. These include the yet-to-be-detected partners of certain known subatomic particles such as quarks and photons. These hypothetical dark matter counterparts are respectively known as *squarks* and *photinos*. There are many theories of and names for these types of dark matter particles, but scientists describe them collectively as *weakly interacting massive particles,* or *WIMPs* (covered later in this chapter).

Taking a Shot in the Dark: Searching for Dark Matter

Around the world, physicists are designing sensitive detectors to find the elusive, telltale signals of dark matter. Some of these detectors analyze the subatomic debris created by giant atom-smashing devices, which briefly recreate the extreme heat, energy, and densities present in the early universe.

The search techniques have to be innovative. After all, scientists are hunting material that by definition we can't see and, aside from exerting a gravitational force, that doesn't interact with other matter.

All methods for detecting and measuring dark matter are indirect, but attempting to understand dark matter isn't a trivial pursuit. As the dominant form of matter in the universe, dark matter profoundly influences the past, present, and future of the universe.

WIMPs: Leaving a weak mark

Consider the effort to find WIMPs (weakly interacting massive particles). No container can trap these weakly interacting particles, but scientists can look for evidence of their existence as they pass through a detector. When a WIMP whizzes past, it slightly heats up one of the detector's atoms, giving it an extra little kick. These encounters, however, are rare. For a typical laboratory detector, this boost may occur only once in many days.

Unfortunately, cosmic rays, which are energetic particles that stream in from all directions of space, can mimic the action of a WIMP. To minimize bombardment from cosmic rays, researchers place the detector in an underground tunnel. Naturally occurring radioactivity from the walls of the tunnel can also heat up the atoms, so the detector is shielded with lead and cooled to near absolute zero to reduce the jiggling of atoms that occurs with increasing vigor at higher temperatures.

MACHOs: Making a brighter image

Because MACHOs aren't microscopic like WIMPS, looking for them is easier. The prime method takes advantage of a mind-bending concept from Einstein's Theory of General Relativity. To wit: Mass distorts the fabric of space and the path of a light wave (as I describe in Chapter 11), which means that an object that by chance lies along the line of sight between Earth and a distant star focuses the light from that star, briefly making it appear brighter. The more massive the object — in this case a MACHO — the brighter the star appears during the alignment.

In effect, the MACHO acts as a miniature gravitational lens, or microlens, bending and brightening the light from the background star. (See Chapter 11 for more on microlensing.)

To search for MACHOs, astronomers have monitored the brightness of stars from one of the Milky Way's nearest neighbors, the Large Magellanic Cloud galaxy. To reach Earth, starlight from the Cloud must pass through the halo of the Milky Way, and MACHOs that reside in the halo should have a measurable effect on that light.

Astronomers have recorded several events in which stars from the Large Magellanic Cloud suddenly brightened and then dimmed again. The number of MACHOs deduced from these observations is nothing much to phone home about, however. Sorry, E.T.

Mapping dark matter with gravitational lensing

On much larger scales, scientists are taking advantage of gravitational lensing to map the dark matter in entire galaxies and even clusters of galaxies.

If a cluster happens to lie in the travel path of light emitted by a background galaxy, it bends and distorts the light — *gravitational lensing* — creating multiple images of the background body. A halo of these ghost images forms around the edges of the cluster as seen from Earth.

To create the exact pattern of observed ghost images, the intervening cluster must have its mass distributed in a particular way. Because most of the cluster's mass consists of dark matter, the process of gravitational lensing reveals how the dark matter is concentrated in the cluster.

Dueling Antimatter: Proving That Opposites Attract

Get ready for another type of matter almost as weird as dark matter, or maybe even weirder. I'm talking about antimatter.

British physicist Paul Dirac predicted the existence of antimatter in 1929. He combined the theories of quantum mechanics, electromagnetism, and relativity in an elegant set of mathematical equations. (If you want to know more about his theories, you'll have to look them up; this isn't a physics book.)

Dirac found that for every subatomic particle, a mirror-image twin should exist, identical in mass but with an opposite electrical charge. Thus the proton has its antiproton and the electron its antielectron.

When a particle and its antiparticle meet, they annihilate each other. Their electric charges cancel out, and their mass is converted into pure energy.

Astronomers have detected antiparticles of the electron and proton in the cosmic rays from deep space. The antielectron is called the *positron* and the antiproton is simply the *antiproton*. Experiments are in progress to search for antihelium in the cosmic rays. Physicists have actually made antiparticles and even entire antiatoms, such as antihydrogen, in the laboratory. Doctors use beams of antiparticles to diagnose and treat cancer.

Astronomers studying gamma rays from space have observed a form of light known as annihilation radiation. Gamma rays are shorter and more energetic than X-rays. When an electron and its antiparticle, the positron, meet, they annihilate, releasing gamma rays of known wavelength. These telltale rays have been detected from several places in our galaxy, including a wide region in the direction toward the center of the Milky Way. Annihilation radiation has also been detected from some very powerful solar flares (see Chapter 10 for more about solar flares).

On the cosmic scale, the big mystery is why the universe contains so many more particles than antiparticles. Experiments are underway to find out why. Presumably, the Big Bang forged equal numbers of both. At least we know we have billions of years to solve the problem before the universe (and us with it!) slips away to whatever fate is in store for it.

Ron Cowen, who covers astronomy and space for *Science News,* originally contributed this chapter. The author, Stephen P. Maran, updated it for this edition of *Astronomy For Dummies.* All opinions expressed in this chapter are those of the author.

Chapter 16

The Big Bang and the Evolution of the Universe

In This Chapter

▶ Evaluating evidence for the Big Bang

▶ Understanding inflation and the expansion of the universe

▶ Delving into dark energy

▶ Examining the cosmic microwave background

▶ Measuring the universe's age

*O*nce upon a time, 13.7 billion years ago, the universe as we know it didn't exist. No matter, no atoms, no light, no photons; not even space or time.

Suddenly, perhaps in an instant, the universe took form as a tiny dense speck filled with light. In a minuscule fraction of a second, all the matter and energy in the cosmos came into being. Much smaller than an atom, the infant universe was searingly hot, a fireball that began mushrooming in size and cooling at a furious rate.

Astronomers and people the world over have come to know this picture of the birth of the universe as the *Big Bang* theory.

The Big Bang wasn't like a bomb that explodes into the environment — there was no environment until the Big Bang occurred — it was the origin and rapid expansion of space itself. During the first trillion-trillion-trillionth of a second, the universe grew more than a trillion-trillion-trillion times bigger. From an original smooth mixture of subatomic particles and radiation arose the collection of galaxies, galaxy clusters, and superclusters present in the universe today. It boggles the mind to think that the largest structures in the universe, congregations of galaxies that stretch hundreds of millions of light-years across the sky, began as subatomic fluctuations in the energy of the infant cosmos. But that's what scientists believe about how the universe took shape.

In this chapter, I cover evidence supporting the Big Bang theory, the expansion of the universe, and related information on dark energy, the cosmic microwave background, the Hubble constant, and standard candles.

For more information on the concepts in this chapter, visit U.C.L.A.'s Frequently Asked Questions in Cosmology site at `www.astro.ucla.edu/~wright/cosmology_faq.html`.

Assessing Evidence for the Big Bang

Why believe that the universe began with a bang?

Astronomers cite three different discoveries that make a compelling case for the theory:

- ✔ **The universe is expanding.** Perhaps the most convincing evidence for the Big Bang comes from a remarkable discovery made by Edwin Hubble in 1929. Up to that time, most scientists viewed the universe as static — stock still and unchanging. But Hubble discovered that the universe is expanding. Groups of galaxies are flying away from each other, like debris flung in all directions from a cosmic explosion, but they haven't been flung apart into space; the space itself between them is expanding, which makes them move farther and farther apart.

    It stands to reason that if galaxies are flying apart, they were once closer together. Tracing the expansion of the universe back in time, astronomers aided by telescopes and by observatories in space found that 13.7 billion years ago (give or take 100 million years), the universe was an incredibly hot, dense place in which a tremendous release of energy triggered an enormous explosion.

- ✔ **The cosmic microwave background.** In the 1940s, physicist George Gamow realized that a Big Bang would produce intense radiation. His colleagues suggested that remnants of this radiation, cooled by the expansion of the universe, may still exist — like the fumes that persist from an extinguished house fire.

    In 1964, Arno Penzias and Robert Wilson of Bell Laboratories were scanning the sky with a radio receiver when they detected a faint, uniform crackling. What the researchers first assumed was static in their receiver turned out to be the faint whisper of radiation left over from the Big Bang. The radiation is a uniform glow of microwave radiation (short radio waves) permeating space. This *cosmic microwave background* has exactly the temperature that astronomers calculate that it should (2.73 K above absolute zero, which is −273.16°C or −459.69°F) if it has cooled steadily since the Big Bang. For their noble discovery, Penzias and

Wilson shared the 1978 Nobel Prize in physics. (See the section "Pulling Universal Info from the Cosmic Microwave Background," later in this chapter, for the full scoop.)

✔ **The cosmic abundance of helium.** Astronomers have found that the amount of helium among all the baryonic matter in the universe is 24 percent by mass (the rest of baryonic matter is almost entirely hydrogen; iron, carbon, oxygen, and all that good stuff put together is just a trace constituent compared to hydrogen and helium). Nuclear reactions inside stars (see Chapter 11) haven't gone on long enough to produce this amount of helium. But the helium we've detected is just the amount that the theory predicts would've been forged in the Big Bang.

As successful as the standard Big Bang theory has proved to be in accounting for observations of the cosmos, the theory is but a starting point for exploring the early universe. For example, the theory, despite its name, doesn't suggest a source for the cosmic dynamite that sparked the Big Bang in the first place.

Inflation: A Swell Time in the Universe

Aside from ignoring the source of the expansion-causing explosion, the Big Bang theory has other shortcomings. In particular, it doesn't explain why regions of the universe that are separated by distances so vast that they can't communicate — even by a messenger traveling at the speed of light — look so similar to each other.

In 1980, physicist Alan Guth devised a theory, which he called *inflation,* that can help explain this puzzle. He suggested that a tiny fraction of a second after the Big Bang, the universe underwent a tremendous growth spurt. In just 10^{-32} seconds (a hundred-millionth of a trillionth of a trillionth of a second), the universe expanded at a rate far greater than at any time in the 13.7 billion years that have elapsed since.

This period of enormous expansion spread tiny regions — which had once been in close contact — out to the far corners of the universe. As a result, the cosmos looks the same, on the large scale, no matter what direction you point a telescope. (Think of a big, lumpy ball of dough: If you roll the dough over and over again with your rolling pin, eventually you smooth all the lumps out and create a uniform sheet of dough.) Indeed, inflation expanded tiny regions of space into volumes far bigger than astronomers can ever observe. This expansion suggests the intriguing possibility that inflation created universes far beyond the scope of our own. Instead of a single universe, a collection of universes, or a *multiverse,* may exist. But I'm adverse to that theory — one universe is hard enough to comprehend as it is!

Inflation had another effect: The infinitesimally short but extraordinarily great growth spurt after the Big Bang captured random, subatomic fluctuations in energy and blew them up to macroscopic proportions. By preserving and amplifying these so-called *quantum fluctuations,* inflation produced regions of the universe with slight variations in density from one to another.

Due to inflation and the quantum fluctuations, some regions of the universe contain more matter and energy, on average, than other regions. As a result, there are cold spots and hot spots in the temperature of the cosmic microwave background (see Figure 16-1). Over time, gravity molded these variations into the spidery networks of galaxy clusters and giant voids that fill our universe today. Check out "Pulling Universal Info from the Cosmic Microwave Background," later in this chapter, for more information.

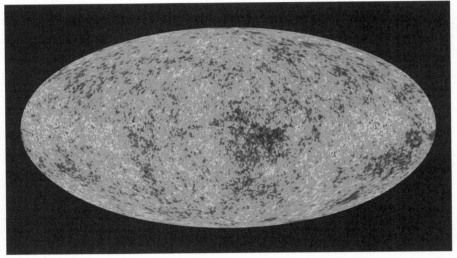

Figure 16-1:
A "baby picture" of the universe from the Wilkinson Microwave Anisotropy Probe satellite.

Courtesy of NASA/WMAP Science Team

The following sections cover two other interesting facets of inflation: the vacuum where inflation gets its power and the relationship between inflation and the universe's shape.

Something from nothing: Inflation and the vacuum

Ironically, the reservoir of energy that powers inflation comes from nothing: the *vacuum.* According to quantum theory, the vacuum of space is far from empty. It seethes with particles and antiparticles that are constantly being created and destroyed. Tapping into this energy, theorists suggest, provided the Big Bang with its explosive energy and radiation.

The vacuum has another bizarre property: It can exert a repulsive gravitational force. Instead of pulling two objects together, *repulsive gravity* pushes them farther apart. Repulsive gravity may have led to the brief but powerful era of inflation.

Just like economic inflation, cosmic inflation generates high interest! And this bubble won't burst.

Falling flat: Inflation and the shape of the universe

The inflation process, at least in its simplest imagined form, would have imposed another condition on the universe: making the geometry of the universe flat. This rapid period of expansion would have stretched out any curvature in the cosmos like a balloon blown to enormous proportions.

For the universe to be flat, it must contain a very specific density, called the *critical density.* If the density of the universe is greater than the critical value, gravity's pull will be strong enough to reverse the expansion, eventually causing the universe to collapse into what astronomers call the *Big Crunch.*

Such a universe would curve back on itself to form a closed space of finite volume, like the surface of a sphere. A starship traveling in a straight line would eventually find itself back where it started. Mathematicians call this geometry *positive curvature.*

If the density is less than the critical value, gravity can never overpower the expansion, and the universe will continue to grow forever. Such a universe has *negative curvature,* with a shape akin to a horse's saddle.

Although inflation theory demands that the universe is flat, several types of observations have revealed that the universe doesn't have enough matter (whether normal or dark matter — see Chapter 15) to attain critical density.

So if the universe is flat, matter as we know it, or even as we don't know it, can't do the trick. But like Mighty Mouse, energy can save the day! In fact, it can save the universe, and recent research shows that it does. The data contained in the "baby picture of the universe" shown in Figure 16-1, which is a sky map of the cosmic microwave background radiation as measured by a NASA satellite called the Wilkinson Microwave Anisotropy Probe, has convinced essentially all cosmologists that the universe is flat and that energy is responsible. But not energy as we ever knew it before; dark energy is the hero. Read on to discover the dark side.

Dark Energy: Stepping on the Universal Accelerator

Dark energy has a startling effect: It exerts a repulsive gravitational force. That's all scientists know about it; we don't know what dark energy is, so we define it by its observable property, the repulsive force. After the Big Bang and inflation, ordinary gravity slowed the expansion of the universe. But as the universe grew bigger and bigger so that the matter was spread apart into more and more space, the slowing effect of gravity became weaker and weaker. After a while (some billions of years), dark energy's repulsive force took over, causing the universe to expand ever faster. Observations from the Hubble and other telescopes revealed this bizarre phenomenon.

The observations that revealed the existence of dark energy by showing that the expansion of the universe is speeding up were of Type Ia supernovas in distant galaxies. (You can read about Type Ia and other supernovas in Chapter 11.) All supernovas are bright enough to be seen in distant galaxies, but the Ia variety have a special property. Astronomers believe that these explosions all have roughly the same intrinsic brightness, like light bulbs of a known wattage (see the section "In a Galaxy Far Away: The Hubble Constant and Standard Candles" later in this chapter).

Because light from a distant galaxy takes hundreds of millions of years or more to reach Earth, observations of that galaxy may show supernovas that erupted when the universe was much younger. If the expansion of the universe had been slowing down ever since the Big Bang, there would be less distance between Earth and the faraway galaxy — and a shorter travel time for light — than if the universe has continued to expand at a fixed speed. Thus, in the case of a slower expansion, a supernova from a distant galaxy should look slightly brighter.

But in 1998, two teams of astronomers found exactly the opposite result: Distant supernovas looked slightly dimmer than expected, as if their home galaxies were farther away than calculated. It appears that the universe has revved up its rate of expansion.

Pulling Universal Info from the Cosmic Microwave Background

The cosmic microwave background (the faint whisper of radiation left over from the Big Bang) represents a snapshot of the universe when it was 379,000 years old. Before that time, a fog of electrons pervaded the infant universe, and radiation created in the Big Bang couldn't stream freely through space. The negatively charged particles repeatedly absorbed and scattered the radiation.

Around the time that the cosmos celebrated its 379,000th birthday, the universe became cool enough for electrons to combine with atomic nuclei, meaning there wasn't an abundance of particles to scatter and absorb radiation. The absorbing fog was lifted. Today we detect the light from the universe at age 379,000 years — now shifted in wavelength by the expansion of the universe — as microwaves and far-infrared light.

Finding the lumps in the cosmic microwave background

When Penzias and Wilson first detected the cosmic microwave background in the 1960s, it appeared to have a perfectly uniform temperature across the sky. No regions in the sky were ever so slightly hotter or colder, at least not to the detection limits of the available instruments. That was a puzzle, because such tiny variations in temperature have to be present to explain how the universe could have begun as a smooth soup of particles and radiation and evolved into a lumpy collection of galaxies, stars, and planets.

According to theory, the infant universe wasn't perfectly smooth. Like lumps in a bowl of porridge, it had slightly overdense and underdense places, with more atoms per cubic inch or fewer atoms per cubic inch, respectively. These places represent the tiny seeds around which matter could have started to clump together to form galaxies. Scientists should now see the variations in density as tiny fluctuations or anisotropies in the temperature of the cosmic microwave background. (An *anisotropy* is a difference in the physical properties of space, such as temperature and density, along one direction from the properties in another direction.)

In 1992, NASA's Cosmic Background Explorer satellite, which just three years earlier had measured the temperature of the microwave background to an unprecedented accuracy, achieved what many astronomers consider an even greater triumph: It detected hot and cold spots in the cosmic microwave background.

The variations are indeed minuscule — less than 10,000th of a degree K colder or hotter than the average temperature of 2.73 K. The princess who could feel a pea through a big stack of mattresses wouldn't have felt these differences. Nonetheless, these cosmic ripples are large enough to account for the growth of structure in the universe. You can sleep on that.

Mapping the universe with the cosmic microwave background

In the search to find out if the universe is flat or saddle-shaped, scientists looked to the cosmic microwave background for answers. A flat universe would dictate that the temperature fluctuations have a particular pattern. A slew of balloon-borne and ground-based telescopes suggested that the microwave background may have this pattern.

In 2003, NASA reported that its Wilkinson Microwave Anisotropy Probe (WMAP) satellite mapped and measured the microwave background over the entire sky in sharper detail than ever before. The WMAP team, led by Dr. Charles L. Bennett, answered most of the existing questions about the Big Bang except what made it happen and what exactly is dark energy. The team found that

✔ The present age of the universe is 13.7 billion years.

✔ The cosmic microwave background originated when the universe was 379,000 years old.

✔ The first stars began to shine about 200 million years after the Big Bang.

✔ The universe is flat, consistent with the theory of inflation (see the section "Inflation: A Swell Time in the Universe" earlier in this chapter).

✔ The relative amounts of mass energy in the universe are as follows:

• Normal matter (baryonic matter like that found on Earth): 4 percent

• Dark matter (see Chapter 15): 23 percent

• Dark energy: 73 percent

Scientists had rough estimates for all these quantities, but now we have precise values.

You can read all about WMAP and its findings at the official Web site of Goddard Space Flight Center, `map.gsfc.nasa.gov`. My favorite part is the animation of quantum fluctuations.

In a Galaxy Far Away: The Hubble Constant and Standard Candles

One of the longest running questions in astronomy used to be "How old is the universe?" Now, thanks to WMAP, the Hubble Space Telescope, and other instruments, we know that the answer is 13.7 billion years.

So how did scientists figure out this magic number? They relied on information connected to the expansion of the universe: the Hubble constant and standard candles, which I cover in the following sections.

The Hubble constant: How fast do galaxies really move?

Cosmic age estimates have depended on a number that has held the attention of astronomers for decades: the *Hubble constant,* which represents the rate at which the universe is currently expanding. The number dates back to 1929, when Edwin Hubble found evidence that we live in an expanding universe. In particular, he made the remarkable discovery that every distant galaxy (those beyond the Local Group of Galaxies, which I describe in Chapter 12) appears to be racing away from our home galaxy, the Milky Way.

Hubble found that the more remote the galaxy, the faster it recedes. This relationship is known as *Hubble's Law.* For example, consider two galaxies, one of which lies twice as far from the Milky Way as the other. The galaxy that resides twice as far away appears to move away twice as fast. (According to Albert Einstein's Theory of General Relativity, the galaxies themselves don't move; rather, the fabric of space in which they reside expands.)

The constant of proportionality that relates the distance of a galaxy to its recession speed is known as the Hubble constant, or H_o. In other words, the speed at which a galaxy recedes is equal to H_o multiplied by the galaxy's distance. H_o thus provides a measure of the rate of universal expansion and, by implication, its age.

The Hubble constant is measured in kilometers per second per megaparsec. (One megaparsec is 3.26 million light-years.) After years of study, astronomers using the Hubble Space Telescope reported a value of 70 for the Hubble constant. That number means that a galaxy about 30 megaparsecs (about 100 million light-years) from Earth speeds away at 2,100 kilometers per second, which is about 1,300 miles per second. WMAP's findings suggest the value is 71; that's darn good agreement.

But, because dark energy makes the universe expand faster, the Hubble constant doesn't stay constant very long; it gets bigger. So the Hubble constant is more of a "Hubble inconstant."

Standard candles: How do scientists measure galaxy distances?

Most strategies for measuring distance require some kind of *standard candle,* the cosmic equivalent of a light bulb of known wattage.

For instance, suppose that you believe you know the true brightness, or *luminosity,* of a particular type of star. Light from a distant source grows dimmer in proportion to the square of the distance, so the apparent brightness of a star of that same type in a distant galaxy indicates how far away the galaxy lies.

Yellowish, pulsating stars known as *Cepheid variables* remain one of the most credible standard candles for estimating the distance to relatively nearby galaxies (see Chapter 12). These youthful stars brighten and dim periodically. In 1912, Henrietta Leavitt of Harvard College Observatory detected that the rapidity with which Cepheids change their brightness is directly linked to their true luminosity. The longer the period, the greater the luminosity.

Type Ia supernovas (see Chapter 11) are another type of standard candle. Because supernovas are much brighter than Cepheids, we can observe them in much more distant galaxies. Recent calculations of the Hubble constant employed both of these candles and got results in good agreement, with each other and with the data from the WMAP satellite. Astronomers now have reliable data on the current expansion rate of the universe, and we know that dark energy is increasing the expansion rate. But the nature of dark energy remains a deep, dark mystery.

Ron Cowen, who covers astronomy and space for *Science News,* originally contributed this chapter. The author, Stephen P. Maran, updated it for this edition of *Astronomy For Dummies.* All opinions expressed in this chapter are those of the author.

Part V
The Part of Tens

In this part . . .

Do you ever find yourself at a social gathering desperately trying to think of something unique and interesting to say? You search your brain for some crowd-grabbing insight to make everyone in the room take notice of your remarkable intelligence. Well, after reading The Part of Tens, you'll be ready for the next lapse in conversation. I offer you 10 strange facts about space that I guarantee will garner interest. And then I fill you in about 10 major mistakes that people in general, and the media in particular, have made, and keep making, on the topic of astronomy.

Chapter 17

Ten Strange Facts about Astronomy and Space

In This Chapter

▶ Discovering the truth about comet tails, Mars rocks, micrometeorites, and the Big Bang on television

▶ Finding out why Pluto's discovery was an accident, sunspots aren't dark, and rain never hits the ground on Venus

▶ Exploring tidal myths, exploding stars, and Earth's uniqueness

Here are some of my favorite facts about astronomy and, in particular, Earth and its solar system. With the following information under your belt, you may be ready to handle the astronomy questions on television quiz shows and inquiries from friends and family.

You Have Tiny Meteorites in Your Hair

Micrometeorites, tiny particles from space visible only through microscopes, are constantly raining down on Earth. Some fall on you whenever you go outdoors. But without the most advanced laboratory equipment and analysis techniques, you can't detect them. They get lost in the great mass of pollen, smog particles, household dust, and (I'm sorry to say) dandruff that resides on the top of your head. (Check out Chapter 4 for the scoop on meteorites of all sizes.)

A Comet's Tail Often Leads the Way

A comet tail isn't like a horse tail, which always trails behind as the horse gallops ahead. A comet tail always points away from the sun. When a comet approaches the sun, its tail or tails stream behind it, but when the comet heads back out into the solar system, the tail leads the way. (See Chapter 4 for more information about comets.)

Earth Is Made of Rare and Unusual Matter

The great majority of all the matter in the universe is so-called *dark matter*, invisible stuff that astronomers haven't yet identified (see Chapter 15). And most ordinary or visible matter is in the form of plasma (hot, electrified gas that makes up normal stars such as the sun) or degenerate matter (in which atoms or even the nuclei within the atoms are crushed together to unimaginable density, as found in white dwarfs and neutron stars; see Chapter 11). You don't find dark matter, degenerate matter, or much plasma on Earth. Compared to the great bulk of the universe, Earth and earthlings are the aliens. (See Chapter 5 for more about Earth's unique properties.)

High Tide Comes on Both Sides of the Earth at the Same Time

Ocean tides on the side of Earth that faces the moon are no higher than tides on the opposite side of Earth at the same time. This may defy common sense, but not physics and mathematical analysis. (The same goes for the smaller ocean tides raised by the sun.) See Chapter 5 for more about the moon.

On Venus, the Rain Never Falls on the Plain

In fact, the constant rain on Venus never falls on anything. It evaporates before it hits the ground, and the rain is pure acid. (The common name for this is *virga;* see Chapter 6 for more about Venus.)

Rocks from Mars Dot the Earth

People have found more than 30 meteorites on Earth that come from the crust of Mars, blasted from that planet by the impacts of much larger objects — perhaps from the asteroid belt (see Chapter 7 for info on asteroids). But these objects are just the Mars rocks that meteorite experts recognized after their discovery. Statistically, many more undiscovered Mars rocks must have fallen into the ocean or landed in out-of-the-way places where they haven't been spotted. (See Chapter 6 to find out more about Mars.)

Pluto Was Discovered from the Predictions of a False Theory

Percival Lowell predicted the existence and approximate location of Pluto. When Clyde Tombaugh surveyed the region, he discovered the planet. But now scientists know that Lowell's theory, which inferred the existence of Pluto from its gravitational effects on the motion of Uranus, was wrong. In fact, Pluto's mass is very low and incapable of producing the "observed" effects. Furthermore, the "gravitational effects" were just errors in measuring the motion of Uranus. (Not enough information was available about Neptune's motion to study it for clues.) The discovery of Pluto took hard work, but as it happened, it was just plain luck. (See Chapter 9 to find out more about Pluto.)

Sunspots Are Not Dark

Almost everyone "knows" that sunspots are "dark" spots on the sun. But in reality, sunspots are simply places where the hot solar gas is slightly cooler than its surroundings (see Chapter 10 for more explanation). The spots look dark compared to their hotter surroundings, but if all you can see is the sunspot, it looks very bright.

A Star in Plain View May Have Exploded, but No One Knows

Eta Carinae is one of the most massive, fiercely shining stars in our galaxy, and astronomers expect it to produce a powerful supernova explosion at any time, if it hasn't already. But because it takes light 9,000 years to travel from Eta Carinae to Earth, an explosion that occurred less than 9,000 years ago isn't visible to us yet. (See Chapter 11 to discover more about the life cycles of stars.)

You May Have Seen the Big Bang on an Old Television

Some of the *snow* (a pattern of interference that looks like little white spots or streaks on old black-and-white television sets) was actually radio waves the TV antenna received from the cosmic microwave background, a glow from the early universe in the aftermath of the Big Bang (see Chapter 16). When this radiation was actually discovered at the Bell Telephone Laboratories, scientists studied many possible causes of the unexpected "noise" in the radio receiver. They even investigated pigeon droppings as a possible cause but later dropped that suggestion.

Chapter 18

Ten Common Errors about Astronomy and Space

. . . .

In This Chapter

▶ Perusing popular misconceptions about astronomy

▶ Correcting mistakes commonly made by the news and entertainment media

. . . .

In daily life — reading the newspaper, watching the evening news, surfing the Web, or talking to friends — you run across many misconceptions about astronomy. In this chapter, I explain the most common of these errors.

"The Light from That Star Took 1,000 Light-Years to Reach Earth"

Many people mistake the light-year for a unit of time on par with units like a day, month, or an ordinary year. But a light-year is a unit of distance, equal to the length that light travels in a vacuum over a period of one year. (See Chapter 1.)

A Freshly Fallen Meteorite Is Still Hot

Actually, freshly fallen meteorites are cold; an icy frost (from contact with moisture in the air) sometimes forms on a frigid stone that has recently landed. When an eyewitness says that he saw a meteorite fall to the ground and he burned his fingers on the rock, the account is probably a hoax. (See Chapter 4 for more information about meteorites.)

Summer Always Comes When Earth Is Closest to the Sun

The belief that summer comes when Earth is closest to the sun is about the most common error of them all, but common sense should tell you that the belief is false. After all, winter occurs in Australia when the United States is experiencing summer. But on any given day, Australia is the same distance from the sun as the United States. (See Chapter 5 for more explanation of our seasons.)

The "Morning Star" Is a Star

The Morning Star isn't a star; it's always a planet. And sometimes two Morning Stars appear at once, such as Mercury and Venus (see Chapter 6). The same idea applies to the Evening Star: You're seeing a planet, and you may see more than one. "Shooting stars" and "falling stars" are misnomers, too. These "stars" are meteors — the flashes of light caused by small meteoroids falling through Earth's atmosphere (see Chapter 4). Many of the "superstars" you see on television may be just flashes in the pan, but they at least get 15 minutes of fame.

If You Vacation in the Asteroid Belt, You'll See Asteroids All Around You

In just about any movie about space travel, you see a scene in which the intrepid pilot skillfully steers the spaceship past hundreds of asteroids that hurtle past in every direction, sometimes coming five at a time. Moviemakers just don't understand the vastness of the solar system, or they ignore it for dramatic purposes. If you stood on an asteroid smack dab in the middle of the main asteroid belt between Mars and Jupiter, you'd be lucky to see more than one or two other asteroids, if any, with the naked eye. (See Chapter 7 for more information about asteroids.)

Nuking a "Killer Asteroid" on a Collision Course for Earth Will Save Us

You come across many common errors about asteroids, and the recent spate of doomsday movies and media reports on "killer asteroids" have provided ample but unfortunate opportunities to reinforce these misunderstandings among the public.

Blowing up an asteroid on a collision course with Earth with an H-bomb would only create smaller and collectively-just-as-dangerous rocks, all still heading for our planet. A better idea would be to attach a rocket motor to gently propel the asteroid just the slightest bit forward or backward in its orbit, steering it so that it doesn't get to the same place in space as Earth at the same time.

Asteroids Are Round, Like Little Planets

A few of the largest asteroids *are* round, but the great majority are irregular blocks of stone or iron. Many are shaped like peanuts or potatoes and are pitted with craters.

The Sun Is an Average Star

You often hear or read statements that the sun is an average star repeated by journalists and published in books written for the general public by astronomers who should know better. In fact, the vast majority of all stars are smaller, dimmer, cooler, and less massive than our sun (see Chapter 10). Be proud of the sun — it's like a kid from the mythical Lake Wobegon, where the children are all "above average."

The Hubble Telescope Gets Up Close and Personal

The Hubble Space Telescope doesn't snap those beautiful pictures by cruising through space until it floats alongside nebulae, star clusters, and galaxies (see Chapter 12). The telescope stays in close orbit around Earth, and it just takes great photos. It does so because it has incredibly well-made optics and orbits far above the parts of Earth's atmosphere that blur our view with telescopes on the ground.

The Big Bang Is Dead

When an astronomer reports a finding that doesn't fit the current understanding of cosmology, members of the media are prone to pronouncing "the Big Bang is dead." (See Chapter 16 for an explanation of the Big Bang.) But astronomers are simply finding differences between the observed expansion of the universe and specific mathematical descriptions of it. The competing theories — including one that fits the newly reported data — are consistent with the Big Bang; they just differ in the details.

Part VI
Appendixes

The 5th Wave   By Rich Tennant

"Paul, turn off your flashlight. There's a real interesting star cluster I'm trying to get a picture of."

In this part . . .

The appendixes in this part offer information to enhance your skywatching experiences for years to come. The first appendix gives you tables showing the approximate locations — at any time from the year 2006 through 2010 — of the four bright planets you can most easily spot: Venus, Mars, Jupiter, and Saturn. The second appendix offers maps to help you find interesting stars. Finally, I include simple definitions for some astronomy terms that you can use as you enjoy your skywatching hobby.

Appendix A

Finding the Planets: 2006 to 2010

*T*he tables in this appendix give, for the years 2006 through 2010, approximate locations of the four bright planets that astronomers most commonly observe: Venus, Mars, Jupiter, and Saturn. You can usually spot these planets with the unaided eye, and after you find them, you can track them for several consecutive months. (Notes about Mercury's movements are also included when the planet is clearly visible.) For each year, I include separate tables for the twilight periods — dawn and dusk — which are the most convenient times for most people to observe the sky. Each table indicates which direction to look to see the planets. The tables are most accurate for middle northern latitudes.

If you follow the moon daily at dusk or dawn, you'll often spot it near one of the five bright planets (Mercury, Venus, Mars, Jupiter, or Saturn) or near one of the several bright stars or noteworthy patterns in the zodiac: the Pleiades, Hyades, or Aldebaran in Taurus; Pollux or Castor in Gemini; Regulus in Leo; Spica in Virgo; Antares in Scorpius; and the Teapot in Sagittarius. (See Chapter 3 for more information about the zodiac.)

As you follow the planets over days, weeks, or months, you'll notice that they drift past one another, as well as past the same zodiacal stars that the moon does. Planet-watching is an activity that can provide enjoyment for a lifetime!

Martin Ratcliffe, a contributing editor for *Astronomy* magazine, supplied the following charts.

2006

Table A-1	Planets at Dusk (About 45 Minutes after Sunset)				
Month	*Venus*	*Mars*	*Jupiter*	*Saturn*	*Planet Happenings*
January	Setting WSW	S	—	ENE	On Jan. 1, Venus is near the crescent Moon. Venus is visible in early January only. On Jan. 27, Saturn is visible all night.
February	—	S	—	E	—
March	—	SW	—	ESE	—
April	—	W	—	S	—
May	—	W	SE	WSW	On May 4, Jupiter is visible all night.
June	—	Low W	S	Low W	On June 17, Saturn and Mars are very close together in Cancer. Also, Mercury is lower down in nearby Gemini. On June 27, the crescent Moon, Mercury, Saturn, and Mars are together very low in the WNW. Add Jupiter to the south, and four planets are visible at once.
July	—	Low W	SSW	Setting NNW	On July 21, Mars is very close to Regulus in Leo, very low in the W. The crescent Moon joins the pair on July 27.
August	—	Low W	SW	—	—
September	—	—	WSW	—	—
October	—	—	Low WSW	—	From Oct. 10 to 21, Mercury and Jupiter are close but very low soon after the sunset in the

Month	Venus	Mars	Jupiter	Saturn	Planet Happenings
					WSW. On Oct. 24, the crescent Moon joins the pair in bright twilight, setting 40 minutes after the sun.
November	—	—	—	—	—
December	Setting WSW	—	—	—	—

Table A-2	Planets at Dawn (About 45 Minutes before Sunrise)				
Month	Venus	Mars	Jupiter	Saturn	Planet Happenings
January	—	—	SSE	W	—
February	SE	—	S	Setting WNW	Venus, Jupiter, and Saturn are visible stretching from E to W in early February. Venus appears as a thin crescent through a telescope.
March	SE	—	SSW	—	Mercury is visible in late March and early April low in the E, to the far lower left of Venus.
April	ESE	—	SW	—	In early April, Venus appears half full through a telescope.
May	E	—	Setting WSW	—	—
June	E	—	—	—	—
July	ENE	—	—	—	—
August	ENE	—	—	Rising ENE	From Aug. 4 to 14, Mercury and Venus are within three degrees of each other. On Aug. 20 and 21, Mercury is near Saturn in bright twilight very low

(continued)

Table A-2 *(continued)*

Month	Venus	Mars	Jupiter	Saturn	Planet Happenings
					in the ENE. On Aug. 22, the extremely thin crescent Moon is next to Mercury and Saturn, with Venus above the moon. On Aug. 26 and 27, Venus and Saturn are in the same telescopic field of view.
September	Rising E	—	—	E	—
October	E	—	—	SE	—
November	—	—	—	S	On Nov. 8, the transit of Mercury across the sun's disk is visible from North America. Following the transit, Mercury is in the morning sky low in the ESE during the last half of November, appearing to the lower left of a crescent Moon on Nov. 18.
December	—	SE	SE	WSW	From Dec. 9 to 11, there is an excellent conjunction of Jupiter, Mars, and Mercury low in the SE. At their closest, all three planets lie within a one-degree circle. On Dec. 18, Jupiter, Mars, and the crescent Moon congregate near Antares in Scorpius, low in the SE.

2007

Table A-3	Planets at Dusk (About 45 Minutes after Sunset)				
Month	*Venus*	*Mars*	*Jupiter*	*Saturn*	*Planet Happenings*
January	SSW	—	—	—	On Jan. 20, the crescent Moon is near Venus in the low SSW. Mercury appears low in the W in the last week of January.
February	Setting W	—	—	E	On Feb. 2, a near full Moon lies close to Saturn. On Feb. 10, Saturn is visible all night. Mercury is visible below Venus in the first two weeks of the month; both planets are low in the WSW.
March	W	—	—	ESE	—
April	WNW	—	—	S	On Apr. 11, Venus is south of the Pleiades star cluster (also known as the Seven Sisters).
May	WNW	—	—	SW	On May 17, the crescent Moon is near Mercury very low in the WNW and below Venus. On May 19, the moon is near Venus.
June	W	—	SE	W	On June 5, Jupiter is visible all night. On June 12, Venus is near the Beehive star cluster in Cancer. On June 18, the crescent Moon lies between Venus and Saturn. On June 30, Saturn and Venus are in the same binocular field of view.

(continued)

Table A-3 *(continued)*

Month	Venus	Mars	Jupiter	Saturn	Planet Happenings
July	W	—	SSE	W	On July 16, the crescent Moon is between Venus and Saturn with Regulus in Leo nearby. Venus is a crescent in telescopes.
August	Setting W	—	S	—	In mid-August, Venus appears half full.
September	—	—	SSW	—	On Sept. 13, Mercury is near the crescent Moon, setting very low in the WSW. On Sept. 17 and 18, the moon is near Jupiter in the SSW.
October	—	—	SW	—	—
November	—	—	SW	—	On Nov. 12, the crescent Moon is below Jupiter in the SW.
December	—	Rising ENE	—	—	On Dec. 24, Mars is visible all night in Gemini and reaches its highest declination north during the month.

Table A-4   Planets at Dawn (About 45 Minutes before Sunrise)

Month	Venus	Mars	Jupiter	Saturn	Planet Happenings
January	—	SE	SE	W	On Jan. 15, Jupiter is north of Antares in Scorpius and the crescent Moon.
February	—	SE	SSE	Setting WNW	—
March	—	ESE	S	—	Mercury makes an appearance in the ESE; it's brighter toward the end of March and early April. On Mar. 16, a crescent Moon lies between Mercury and Mars.

Month	Venus	Mars	Jupiter	Saturn	Planet Happenings
April	—	ESE	SSW	—	Mars remains in the ESE for some months but increases in altitude each month.
May	—	ESE	SW	—	—
June	—	ESE	Setting WSW	—	—
July	—	ESE	—	—	Mercury is visible in late July and early August low in the ENE. Mars stands high above near the Pleiades.
August	Rising E	ESE	—	—	From Aug. 1 to 3, Mercury is visible low in the ENE near Castor and Pollux in Gemini, and it soon sinks out of view. In late August, Venus begins its morning visibility low in the E, and Mars is very high in Taurus.
September	E	SE	—	Rising E	Saturn is near Regulus in Leo.
October	ESE	SW	—	ESE	On Oct. 15, Venus and Saturn are two degrees apart in Leo. Venus shows its half phase through a telescope at the end of October.
November	ESE	SW	—	SE	Four planets are visible simultaneously when Mercury emerges; Mercury, Venus, Saturn, and Mars span from E to W. On Nov. 7, Mercury is near Spica in Virgo and the crescent Moon. Venus and Saturn lie high above the moon.

(continued)

Table A-4 (continued)

Month	Venus	Mars	Jupiter	Saturn	Planet Happenings
December	SE	W	—	S	On Dec. 5, Venus, the crescent Moon, and Spica in Virgo form an attractive triangle.

2008

Table A-5   Planets at Dusk (About 45 Minutes after Sunset)

Month	Venus	Mars	Jupiter	Saturn	Planet Happenings
January	—	E	—	—	Mercury is in view from Jan. 7 to 31. On Jan. 9, the crescent Moon is near Mercury.
February	—	SE	—	Rising ENE	On Feb. 20, there is a total eclipse of the moon with Saturn four degrees away in Leo. On Feb. 24, Saturn is visible all night.
March	—	SSW	—	E	—
April	—	SSW	—	SSE	From the end of April until mid-May, Mercury makes a very favorable appearance low in the W to WNW.
May	—	W	—	SSW	On May 2, Mercury lies south of the Pleiades. On May 6, the crescent Moon lies near Mercury. On May 22, Mars is closest to the Beehive star cluster in Cancer.
June	—	W	Rising SE	W	Jupiter appears in the SE in the last week of June.

Month	Venus	Mars	Jupiter	Saturn	Planet Happenings
July	—	W	Rising SE	W	On July 6, the crescent Moon is near Mars and Saturn. On July 9, Jupiter is visible all night. On July 10, Mars and Saturn are close to each other in Leo, near Regulus.
August	Setting W	Setting W	SSE	Setting W	Four planets are in the western sky. On Aug. 14, Mercury, Venus, and Saturn are within the same binocular field of view extremely low in the W setting soon after sunset, with Mars nearby to the E of Venus.
September	Setting W	Setting W	S	—	On Sept. 1, Mercury, Venus, and Mars, with the crescent Moon below them, are in the west, setting soon after the sun. On Sept. 11, Mars and Venus are closest.
October	Setting SW	—	S	—	—
November	SW	—	SSW	—	The two brightest planets, Venus and Jupiter, are together in Sagittarius all month; they'll be closest on Nov. 30 with the crescent Moon nearby.
December	SW	—	SW	—	On Dec. 1, the crescent Moon, Venus, and Jupiter are close together in the SW in this year's most spectacular conjunction. On Dec. 29, Mercury and Jupiter set together low in the SW, with the crescent Moon above them. Mercury and Jupiter are closest on Dec. 30 and 31, and the crescent Moon is next to a brilliant Venus.

Table A-6		Planets at Dawn (About 45 Minutes before Sunrise)			
Month	*Venus*	*Mars*	*Jupiter*	*Saturn*	*Planet Happenings*
January	SE	Setting WNW	SE	WSW	Venus and Jupiter grow closer in Sagittarius throughout late January.
February	SE	—	SE	W	On Feb. 1, Jupiter and Venus are in close conjunction, only one moon-width apart and visible in the same low-power field of view. On Feb. 4, the crescent Moon lies south of the bright planets. On Feb. 26, Venus and Mercury are near each other in bright twilight, very low in the SE.
March	Rising ESE	—	SE	Setting WNW	On Mar. 5, the crescent Moon, Venus, and Mercury lie close in the ESE. Through a telescope, Venus is nearly full and Mercury is a gibbous disk. The daylight occultation of Venus by the moon is visible from the United States.
April	Rising E	—	SSE	—	Venus moves out of view in late April.
May	—	—	S	—	—
June	—	—	WSW	—	—
July	—	—	Setting WSW	—	Mercury is visible low in the SE for the first half of July. It telescopically appears as a crescent in the first week of July.
August	—	—	—	—	—
September	—	—	—	Rising E	Saturn appears late in September, rising in the early dawn.

Month	Venus	Mars	Jupiter	Saturn	Planet Happenings
October	—	—	—	ESE	Mercury makes a favorable appearance during the last half of October and early November, passing north of Spica in Virgo on Oct. 31. Saturn stands high above Mercury.
November	—	—	—	SE	—
December	—	—	—	S	—

2009

Table A-7	Planets at Dusk (About 45 Minutes after Sunset)				
Month	Venus	Mars	Jupiter	Saturn	Planet Happenings
January	SW	—	—	—	—
February	WSW	—	—	—	—
March	Setting W	—	—	E	On Mar. 8, Saturn is visible all night.
April	—	—	—	SE	Mercury is best viewed in the evening sky from mid-April to early May. On Apr. 26, Mercury and the moon are near the Pleiades in the WNW (an excellent binocular sight). On Apr. 30, Mercury is closest to the Pleiades low in the WNW.
May	—	—	—	S	—
June	—	—	—	WSW	—
July	—	—	Rising ESE	W	—

(continued)

Table A-7 (continued)

Month	Venus	Mars	Jupiter	Saturn	Planet Happenings
August	—	—	SE	Setting W	On Aug. 14, Jupiter is visible all night. On Aug. 17, Mercury and Saturn are close together, setting low in the W. On Aug. 22, the crescent Moon, Mercury, and Saturn lie along the western horizon low in the W.
September	—	—	SE	—	—
October	—	—	SSE	—	—
November	—	—	S	—	—
December	—	—	SSW	—	Mercury makes a brief appearance in the low SW. On Dec. 18, Mercury lies below the crescent Moon.

Table A-8   Planets at Dawn (About 45 Minutes before Sunrise)

Month	Venus	Mars	Jupiter	Saturn	Planet Happenings
January	—	—	—	SW	—
February	—	Rising ESE	Rising ESE	W	On Feb. 22, the crescent Moon, Mars, Jupiter, and Mercury are in line in the low SE. Binoculars are required for Mars in bright twilight. On Feb. 24, Jupiter and Mercury are closest to each other.
March	—	Rising ESE	ESE	W	On Mar. 1, Mercury and Mars are in the same telescopic field of view, very low in ESE. Mars is faint.

Month	Venus	Mars	Jupiter	Saturn	Planet Happenings
April	Rising E	Rising E	SE	—	On Apr. 22, the crescent Moon occults a crescent Venus as seen from North America (during daylight on the U.S. east coast and darkness on the west coast). Mars lies four degrees south of the moon.
May	E	E	SE	—	On May 21, Venus, Mars, and the crescent Moon form a triangle.
June	E	E	S	—	On June 19, Venus, Mars, and the crescent Moon are in line, with Venus and Mars fewer than two degrees apart. Mercury lies low in the ENE and Jupiter is to the south, offering four visible planets through early July.
July	E	E	SSW	—	On July 13, Venus and Mars are in Taurus. Venus lies north of Aldebaran, and Mars lies south of the Pleiades star cluster. On July 18, the crescent Moon joins them.
August	E	E	Setting WSW	—	—
September	E	ESE	—	—	On Sept. 1, Venus lies near the Beehive cluster in Cancer. On Sept. 20, Venus is near Regulus in Leo.
October	E	SE	—	E	On Oct. 8, Mercury and Saturn are less than one moon-width apart; you can view them in the same telescopic field of view.

(continued)

Table A-8 *(continued)*

Month	Venus	Mars	Jupiter	Saturn	Planet Happenings
					On Oct. 10, four planets are visible, with Mercury, Venus, and Saturn along a six-degree span in the east, while Mars stands high in Gemini to the SE. On Oct. 13, Venus and Saturn are one moon-width apart. On Oct. 16, the crescent Moon joins the Venus-Saturn pairing. On Oct. 24, Mercury is north of Spica in Virgo.
November	ESE	S	—	ESE	On Nov. 1, Mars is near the Beehive star cluster in Cancer. On Nov. 2, Venus is north of Spica in Virgo.
December	—	WSW	—	S	—

2010

Table A-9		Planets at Dusk (About 45 Minutes after Sunset)			
Month	Venus	Mars	Jupiter	Saturn	Planet Happenings
January	—	Rising ENE	SW	—	On Jan. 29, Mars is visible all night.
February	Setting W	E	Setting W	—	On Feb. 16, Jupiter and Venus are one moon-width apart, setting 40 minutes after sunset.
March	Setting W	SE	—	Rising E	Four planets are visible in late March: Mercury, Venus, Mars, and Saturn. On Mar. 22, Saturn is visible all night.

Month	Venus	Mars	Jupiter	Saturn	Planet Happenings
April	W	S	—	ESE	In the first week of April, Mercury and Venus lie together low in the W. On Apr. 14, Mars is north of the Beehive cluster in Cancer. On Apr. 15, the crescent Moon is near Venus in the WNW.
May	WNW	SW	—	SSE	On May 15 and 16, the crescent Moon is near Venus in the WNW.
June	WNW	WSW	—	SW	On June 6, Mars is near Regulus in Leo. On June 14, the crescent Moon is near Venus in the WNW.
July	W	W	—	W	Venus, Mars, and Saturn become closer. On July 27, four planets are visible in the W as Mercury lies close to Regulus in Leo. On July 31, Mars and Saturn are at their closest to each other.
August	W	W	—	Setting W	During the first two weeks of August, Venus, Mars, and Saturn form a nice triangle in Virgo low in the W. On Aug. 12 and 13, the crescent Moon joins the three planets low in the W. On Aug. 31, Venus and Mars are near Spica in Virgo with Saturn in the W.
September	WSW	WSW	Rising E	—	On Sept. 10, the moon, Venus, Mars, and Spica are within an eight-degree circle. On Sept. 21, Jupiter is visible all night.

(continued)

Table A-9 *(continued)*

Month	Venus	Mars	Jupiter	Saturn	Planet Happenings
October	—	Setting WSW	ESE	—	On Oct. 9, Mars appears above the crescent Moon; binoculars are needed to spot Mars.
November	—	Setting WSW	SE	—	On Nov. 7, Mars is north of Antares in Scorpius with the crescent Moon between them; binoculars are required to see them.
December	—	Very low and faint in the WSW (see Planet Happenings)	S	—	On Dec. 6, the moon occults Mars in bright twilight right after sunset on the U.S. east coast (a telescope is required to see Mars in December). This occurs in daylight for the western half of the U.S. Mercury stands to the upper left of the moon.

Table A-10   Planets at Dawn (About 45 Minutes before Sunrise)

Month	Venus	Mars	Jupiter	Saturn	Planet Happenings
January	—	W	—	SW	Mercury is visible rising in the ESE from mid-January to mid-February. On Jan. 13, Mercury is near the crescent Moon.
February	—	Setting WNW	—	WSW	Mercury is rising low in the ESE. On Feb. 11, Mercury is near the crescent Moon.
March	—	—	—	WSW	—
April	—	—	E	—	On Apr. 11, Jupiter appears below the crescent Moon.

Month	Venus	Mars	Jupiter	Saturn	Planet Happenings
May	—	—	ESE	—	On May 9, the crescent Moon is near Jupiter. Mercury is visible low in the E during late May and early June.
June	—	—	SE	—	On June 10, Mercury lies below the crescent Moon.
July	—	—	S	—	—
August	—	—	SW	—	—
September	—	—	WSW	—	Mercury makes a favorable appearance in late September.
October	—	—	—	Rising E	—
November	ESE	—	—	ESE	Venus is near Spica in Virgo for most of November. Saturn stands higher in the same constellation.
December	SE	—	—	SE	On Dec. 2, Venus, Spica in Virgo, and the crescent Moon form a nice triangle, with Saturn standing above them.

Appendix B

Star Maps

The following pages contain eight star maps — four each of the Northern and Southern Hemispheres — to help start you on your starry way.

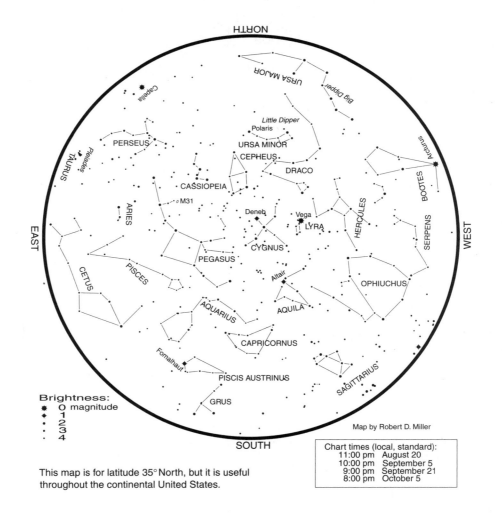

Map by Robert D. Miller

Brightness:
- 0 magnitude
- 1
- 2
- 3
- 4

Chart times (local, standard):
- 11:00 pm   August 20
- 10:00 pm   September 5
- 9:00 pm   September 21
- 8:00 pm   October 5

This map is for latitude 35° North, but it is useful throughout the continental United States.

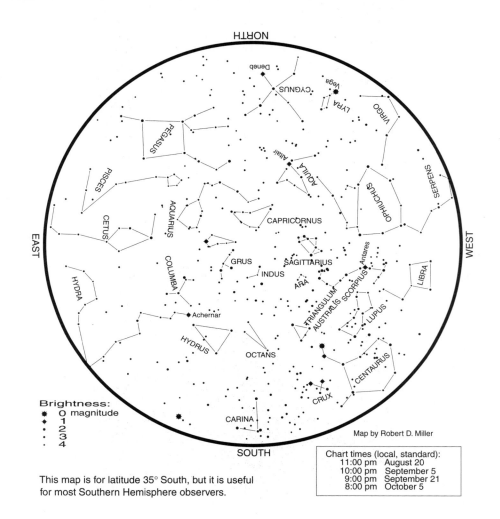

Map by Robert D. Miller

Brightness:
* 0 magnitude
♦ 1
• 2
· 3
· 4

Chart times (local, standard):
11:00 pm   August 20
10:00 pm   September 5
9:00 pm   September 21
8:00 pm   October 5

This map is for latitude 35° South, but it is useful for most Southern Hemisphere observers.

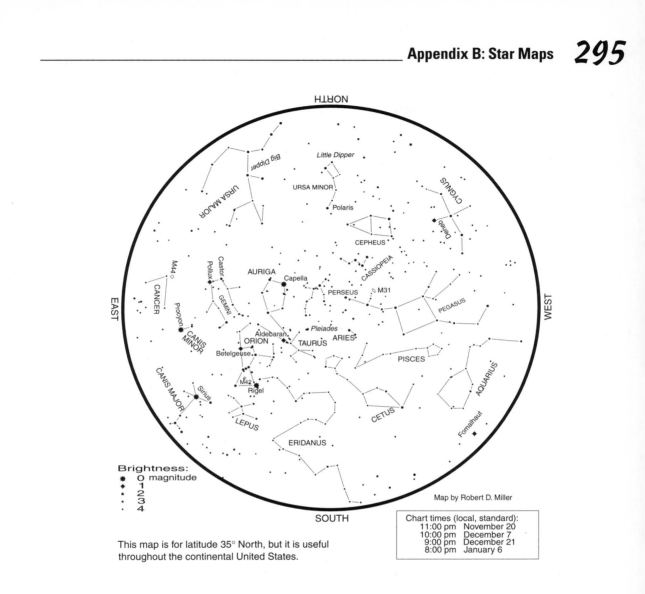

Map by Robert D. Miller

This map is for latitude 35° North, but it is useful throughout the continental United States.

Chart times (local, standard):
11:00 pm   November 20
10:00 pm   December 7
9:00 pm   December 21
8:00 pm   January 6

Brightness:
0 magnitude
1
2
3
4

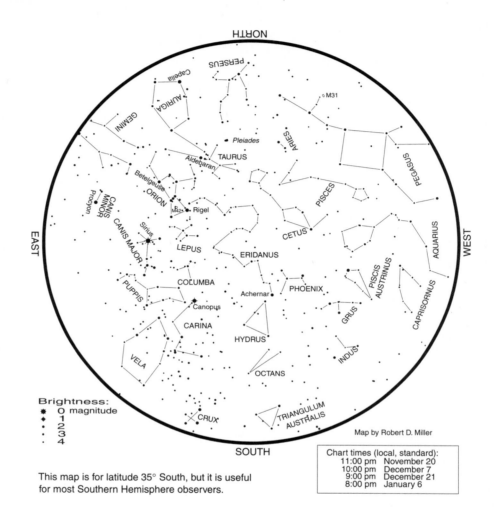

Brightness:
- ✸ 0 magnitude
- ✦ 1
- ● 2
- • 3
- · 4

Map by Robert D. Miller

This map is for latitude 35° South, but it is useful for most Southern Hemisphere observers.

Chart times (local, standard):
11:00 pm	November 20
10:00 pm	December 7
9:00 pm	December 21
8:00 pm	January 6

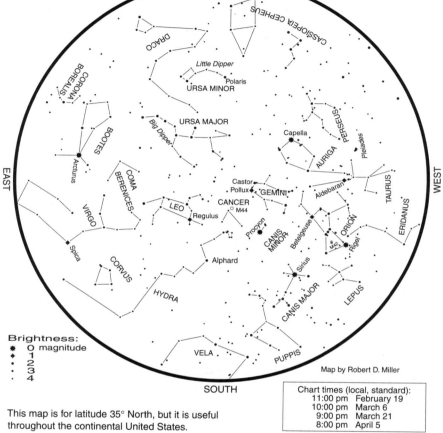

Brightness:
* 0 magnitude
* 1
* 2
* 3
* 4

This map is for latitude 35° North, but it is useful throughout the continental United States.

Map by Robert D. Miller

Chart times (local, standard):
11:00 pm February 19
10:00 pm March 6
9:00 pm March 21
8:00 pm April 5

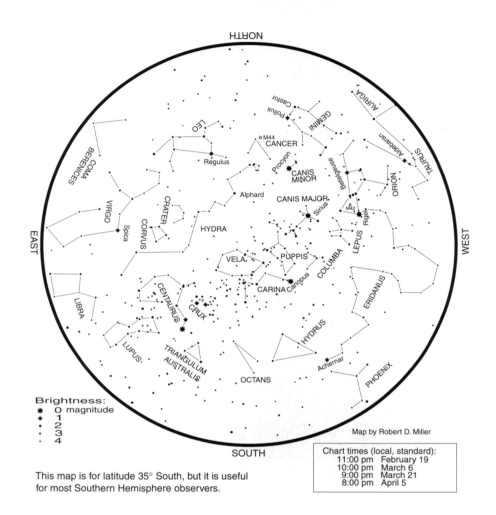

NORTH

AURIGA

Castor
Pollux
GEMINI
Aldebaran
TAURUS

CANCER
o M44

LEO
Betelgeuse
ORION

Regulus
Procyon
CANIS
MINOR

M42
Rigel

COMA
BERENICES

Alphard
CANIS MAJOR.
Sirius

EAST

VIRGO
CRATER
HYDRA

WEST

Spica
CORVUS
PUPPIS
LEPUS

VELA.
COLUMBA

LIBRA
CENTAURUS
CARINA Canopus
ERIDANUS

CRUX

LUPUS.
TRIANGULUM
AUSTRALIS
HYDRUS

OCTANS
Achernar
PHOENIX

Map by Robert D. Miller

Brightness:
✳ 0 magnitude
◆ 1
♦ 2
• 3
· 4

SOUTH

This map is for latitude 35° South, but it is useful
for most Southern Hemisphere observers.

Chart times (local, standard):
11:00 pm   February 19
10:00 pm   March 6
9:00 pm   March 21
8:00 pm   April 5

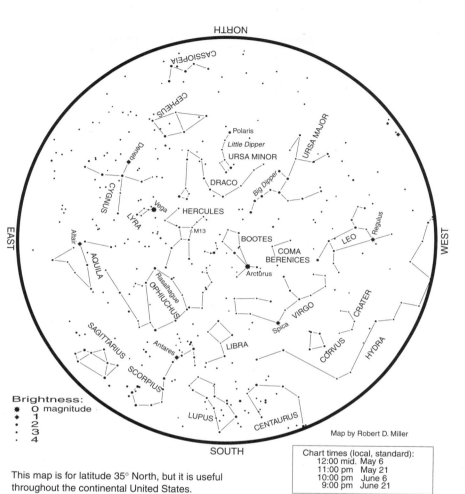

This map is for latitude 35° North, but it is useful throughout the continental United States.

Brightness:
- 0 magnitude
- 1
- 2
- 3
- 4

Map by Robert D. Miller

Chart times (local, standard):
12:00 mid. May 6
11:00 pm May 21
10:00 pm June 6
9:00 pm June 21

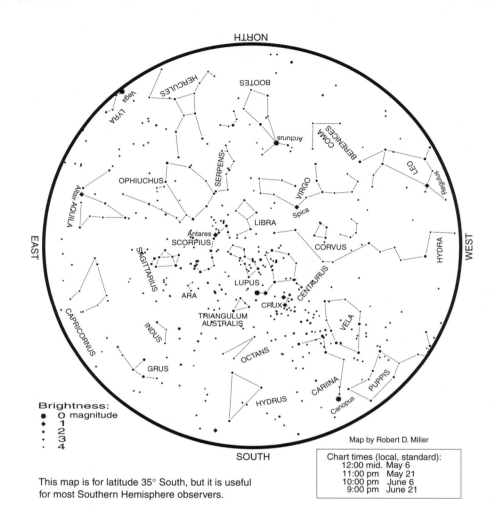

Brightness:
* 0 magnitude
* 1
* 2
* 3
* 4

Map by Robert D. Miller

Chart times (local, standard):
12:00 mid. May 6
11:00 pm May 21
10:00 pm June 6
9:00 pm June 21

This map is for latitude 35° South, but it is useful for most Southern Hemisphere observers.

Appendix C

Glossary

antimatter: Matter composed of antiparticles that have the same mass but opposite electrical charge of ordinary particles.

asterism: A named pattern of stars, such as the Big Dipper, that isn't one of the 88 official constellations.

asteroid: One of many small, rocky, and/or metallic bodies orbiting the sun.

binary star: Two stars orbiting around a common center of mass in space; also called a binary system.

black hole: An object with a gravity so strong that nothing inside can escape, not even a ray of light.

bolide: A very bright meteor that appears to explode or that produces a loud noise.

comet: One of many small bodies made of icy and dusty matter orbiting the sun.

constellation: Any one of 88 regions on the sky, typically named after an animal, object, or ancient deity; for example, Ursa Major, the Great Bear.

crater: A round depression on the surface of a planet, moon, or asteroid created by the impact of a falling body, a volcanic eruption, or the collapse of an area.

dark energy: An unexplained physical process that acts as if it was repulsive gravitational force, causing the universe to expand at a greater and greater rate.

dark matter: An unknown substance in space that has a gravitational effect on celestial objects, which is how astronomers detect its existence.

Doppler Effect: The process by which light or sound is altered in perceived frequency or wavelength by the motion of its source with respect to the observer.

double star: Two stars that appear very close to each other on the sky and that may be physically associated (a binary star) or may be unrelated to each other and at different distances from Earth.

eclipse: The partial disappearance (partial eclipse) or total disappearance (total eclipse) of a celestial body when another object passes in front of it or when it moves into the shadow of another object.

ecliptic: The apparent path of the sun across the background of the constellations.

fireball: A very bright meteor.

galaxy: A huge system of billions of stars, sometimes with vast amounts of gas and dust.

gamma ray burst: An intense outburst of gamma rays that comes without warning from a random spot in the distant universe.

meteor: The flash of light caused by the fall of a meteoroid through Earth's atmosphere; often incorrectly used to mean the meteoroid itself.

meteorite: A meteoroid that landed on the earth.

meteoroid: A rock in space, composed of stone and/or metal; probably a chip from an asteroid.

Near Earth Object: An asteroid or comet that follows an orbit that brings it close to Earth's orbit around the sun.

nebula: A cloud of gas and dust in space that may emit, reflect, and/or absorb light.

neutrino: A subatomic particle that has no electric charge and an extremely small mass. It can pass through a whole planet or even the sun.

neutron star: An object only tens of miles across but greater in mass than the sun (all pulsars are neutron stars, but not all neutron stars are pulsars).

occultation: The process by which one celestial body passes in front of another, blocking it from the view of an observer.

orbit: The path followed by a celestial body or a spacecraft.

planet: A large, round object that forms in a flattened cloud around a star and which — unlike a star — doesn't generate energy by nuclear reactions.

planetary nebula: A glowing, expanding gas cloud that was expelled in the death throes of a sun-like star.

pulsar: A fast-spinning, tiny, and immensely dense object that emits light, radio waves, and/or X-rays in one or more beams like the beam from a lighthouse.

quasar: A small, extremely bright object at the center of a distant galaxy, thought to represent the emission of much energy from the surroundings of a giant black hole.

red giant: A large, very bright star with a low surface temperature; also a late stage in the life of a sun-like star.

redshift: An increase in the wavelength of light or sound, often due to the Doppler Effect or, in the case of distant galaxies, the expansion of the universe.

rotation: The spinning of an object around an axis that passes through it.

seeing: A measure of the steadiness of the air at a place of astronomical observation (when the seeing is good, the images you view through telescopes are sharper).

SETI: The Search for Extraterrestrial Intelligence, a program of radio astronomy observations (and other observations) that seeks to detect messages from intelligent civilizations elsewhere in space.

solar activity: Changes in the appearance of (and in the radiation from) the sun that occur from second to second, minute to minute, hour to hour, and even year to year. It includes solar events such as solar flares and coronal mass ejections and solar features such as sunspots.

spectral type: A classification applied to a star based on the appearance of its spectrum, usually related to the temperature in the region where the visible light from the star originates.

star: A large mass of hot gas held together by its own gravity and fueled by nuclear reactions.

star cluster: A group of stars held together by their mutual gravitational attraction that formed together at about the same time (types include globular clusters and open clusters).

supernova: An immense explosion that disrupts an entire star and that may form a black hole or a neutron star.

terminator: The line separating the illuminated and dark parts of a celestial body that shines by reflected light.

transit: The movement of a smaller object, such as Mercury, in front of a larger object, such as the sun.

variable star: A star that changes perceptibly in brightness.

white dwarf: A small, dense object shining from stored heat and thus fading away; the final stage in the life of a sun-like star.

zenith: The point on the sky that's directly above the observer.

Sky Measures

arc minutes/arc seconds: Units of measurement on the sky. A full circle around the sky consists of 360 degrees, each divided into 60 arc minutes; each arc minute is divided into 60 arc seconds.

astronomical unit (A.U.): A measure of distance in space, equal to the average distance between Earth and the sun — about 93 million miles.

declination: On the sky, the coordinate that corresponds to latitude on Earth and that's measured in degrees north or south of the Celestial Equator.

light-year: The distance light travels in a vacuum (for example, through space) in one year; about 5.9 trillion miles.

magnitude: A measure of the relative brightness of stars, with smaller magnitudes corresponding to brighter stars. For example, a 1st magnitude star is 100 times brighter than a 6th magnitude star.

right ascension: On the sky, a coordinate that corresponds to longitude on Earth and that's measured eastward from the vernal equinox (a point on the sky where the celestial equator crosses the ecliptic and where the sun is located on the first day of spring in the Northern Hemisphere).

Index

• Symbols •

' (arc minute), measuring with, 109, 304
" (arc second), measuring with, 109, 304

• A •

AAVSO (American Association of Variable
    Star Observers), 195–196
absolute magnitude, 47
accretion disk, black holes and, 223
active galactic nuclei (AGN)
  energy powering, 229
  terminology for, 227–229
  unified model of, 230
albedo map, viewing Mars with, 112
Algol, the Demon Star, eclipsing, 192–193
aliens. *See* Search for Extraterrestrial
    Intelligence (SETI)
Allen Telescope Array, 239
Alpha Canis Majoris, 194–195
Alpha Centauri system, 193–194
ALPO (Association of Lunar and Planetary
    Observers), 113
alt-azimuth mount, 52, 53
American Association of Variable Star
    Observers (AAVSO), 195–196
Andromeda Galaxy
  distance of, 22
  near Milky Way, 201
  viewing, 214–215
anisotropies, in cosmic microwave
    background, 259
annihilation radiation, 251
antimatter, 251–252, 301
antipode of Caloris, 98
apparent magnitude, 47
arc aurora, 79
arc minute ('), measuring with, 109, 304
arc second ("), measuring with, 109, 304

artificial satellites
  about, 72–73
  around Mars, 100
  viewing, 73–74
Association of Lunar and Planetary
    Observers (ALPO), 113
associations, of Milky Way, 199
asterism, defined, 10, 301
Asteroid Belt, 116–117, 270
asteroidal meteoroid, 58
asteroids
  defined, 115–116, 301
  invisibility of multiple, 270
  largest of Asteroid Belt, 116
  naming, 71
  protecting Earth from, 119–120, 270–271
  shape of, 271
  threat of near-Earth, 118–119
  viewing, 120–121
astronomers, function of, 10–11
Astronomical League, 31
astronomical observations, reporting, 83
Astronomical Society of the Pacific, 31
astronomical unit (A.U.), defined, 23, 304
astronomy clubs, 30–31, 56
astronomy, defined, 10
Astronomy magazine, 32
AstroWeb Directory of Observatories and
    Telescopes, 36
atmosphere
  on Earth, 80
  on Jupiter and Saturn, 124
  on Mercury, 99
  on moon, 93
atmospheric turbulence, good versus bad
    seeing and, 55
A.U. (astronomical unit), defined, 23, 304
aurora australis (Southern Lights), 79
aurora borealis (Northern Lights), 79
auroral ovals, 79

auroral zones, 79
auroras, 79
autumnal equinox, 85
axis, tilt of, cause of seasons, 84–86

• *B* •

Baily's Beads, during eclipse, 165
Baltis Vallis (rille), 99–100
barometric pressure, 99, 124
barred spiral galaxies, 212
baryonic dark matter, 248
Beehive star cluster, 202
Beta Persei, eclipsing, 192–193
Betelgeuse (star), in Orion, 44–45, 195
Big Bang
    about, 253–254
    defined, 27
    evidence of, 254–255
    reliability of theory, 271–272
    TV evidence of, 268
Big Dipper, 11, 44
binary stars
    about, 183–184
    defined, 301
    Doppler Effect and, 184–185
    eclipsing, 192–193
binoculars
    about, 48
    acquiring basic, 56
    H-alpha, 158
    purpose of, 41–42
    selecting, 49–50
    sun safety and, 54
    understanding numbers on, 49
biosphere, 80
BL Lacs (BL Lacertae objects), 228
black holes
    about, 176–178
    defined, 219, 301
    escaping from, 219–220
    galactic bulge and, 230
    measurements of, 222
    observable material around, 223
    parts of, 220–223
    powering AGN, 229
    types of, 220
    warping space and time, 224–225

blazars, 228
bolide (meteor), defined, 60, 301
brilliance, measuring star, 20–22
British Astronomical Association, 31
brown dwarf stars, on H-R diagram, 183
Butler, Paul (extrasolar planet studies), 243

• *C* •

Callisto, 127
Caloris basin, 98
canals, on Mars, 102
canyons, on Mars, 102
Carina Nebula, 210
Cassegrain reflector telescopes, 51
Cassini division, 130
Cassini probe, Saturn and, 132, 133
cataclysmic variable, 190
celestial sphere, understanding, 25
Centaurus A Galaxy, 216
central peaks, on moon, 91
central stars of planetary nebulae, 173
Cepheid variable stars, 188–189
Ceres (asteroid), 116, 117
Charon (Pluto's moon), 138–139
Chicxulub crater, 118
Chiu, Dr. Hong-yee (defining quasars), 225
chromosphere (sun), 149
Ciel & Espace magazine, 33
circumpolar stars, 44
clouds, 123, 124–125
Coal Sack nebula, 210
cold dark matter, 247
color, 53, 178–180
color sphere (sun), 149
color-magnitude diagram. *See* H-R diagram
coma, 66, 68
cometary meteoroid, 58
comets
    defined, 65, 301
    finding with ephemeris, 26
    hitting Jupiter, 129
    versus meteors, 65
    naming, 71
    random searching for, 70
    reporting, 72
    searching for, 70
    structure of, 65–68

systematic searching for, 70–72

tails, 67–68, 265

viewing, 68–69

comparison charts, 195

comparison stars, 41

conjunction, 106, 107

constellations. *See also specific
    constellations*

brightest stars of, 16–19

containing brightest stars, 46

defined, 301

naming, 12–16

tracking, 43

continental drift, 78, 100

convection zone (sun), 148

Coordinated Universal Time (UTC), 83, 84

Copernicus, Nicholas (17th-century Polish
    astronomer), 113

core (sun), 147–148

corona (sun), 149, 163

coronal mass ejections, 150–151, 153

cosmic microwave background

Big Bang theory and, 254–255

finding variations in, 259

mapping universe with, 260

TV snow and, 268

understanding universe from, 258–259

cosmic voids, super clusters and, 218

Crab Nebula, 209

craters

on asteroids, 117

defined, 301

on Mercury, 98

on moon, 90–91

on Venus, 100

critical density, 248, 257

cruises, eclipse, 38, 39–40

cryosphere, 80

cryovolcanism, 137

curtain aurora, 79

• D •

dark energy, 258, 301

dark matter

about, 245–246

classes of, 248–249

defined, 301

evidence of, 246–248

searching for, 249–251

dark nebulae, 205–206

declination circles, 25

declination (Dec), 24–26, 106, 304

degrees, measuring sky objects with, 109

Delta Aquarids (meteor showers), 62

dimming technique, 241

Dione (Saturn's moon), 133

Dirac, Paul (antimatter studies), 251

distance

measuring with light-years, 22–23, 269

measuring with standard candles, 262

Dobsonian reflector telescope, 51, 52

dog star (Sirius), 194–195

Doppler Effect

binary stars and, 184–185

defined, 301

measuring binary star eclipses, 192

Double cluster, 202

double stars, 187, 302

Drake Equation, 234, 236

Drake, Frank (alien searches), 234

drapery aurora, 79

Dubhe (star), finding North Star with, 44

Dumbbell Nebula, 209

dust tails, comet, 67

dwarf galaxies, 213–214

• E •

Earth

about, 77–78

asteroids hitting, 117

estimating age of, 86

four views of, 80

geomagnetic field on, 81

magnetic properties of seafloor, 82

Mars rocks on, 266–267

versus Mercury, Venus, Mars, 103–104

motion of, 27–28

ocean tides on, 266

as origin of moon, 94–95

phases of, 89

protecting from asteroids, 119–120

threat of asteroids near, 118–119

understanding regions of, 80–81

unique characteristics of, 78–79

unusual matter of, 266

eclipse cruises and tours, 38–40
Eclipse Shades, 164
eclipses
   defined, 302
   lunar, 89–90
   path of totality, 38
   solar, 162–166
   viewing Jupiter's moons and, 128
eclipsing binary stars, 192–193
ecliptic, sun's movement and, 43
Edmund Scientific, moon maps at, 92
Eight-Burst Nebula, 210
Einstein, Albert (theory of gravity), 26, 27
elliptic latitude, 106
elliptic longitude, 106
elliptical galaxies, 212–213
elongation, 106
end states of stellar evolution
   black holes, 176–178
   central stars of planetary nebulae, 173
   neutron stars, 175–176
   star lifecycle and, 170
   supernovas, 174–175
   white dwarf stars, 174
energy
   dark, 258
   generated on Jupiter and Saturn, 124
   powering AGNs, 229
   powering inflation, 256–257
   produced by sun, 152
   sun's nuclear fusion and, 146
ephemeris, finding comets with, 26
equatorial mount, 52–53
equipment, acquiring basic, 56. *See also
   specific equipment*
escape velocity, 219–220
Eta Aquarids (meteor showers), 62
Europa, 128
European Space Agency, Mars images
   from, 100
Evening Star, changing, 105
event horizon, black holes, 177, 221–222
expanding universe. *See* universe
extrasolar planets
   around 51 Pegasi, 242–243
   around Upsilon Andromedae, 243
   continuing search for, 243–244
   finding, 240–241
extrinsic variable stars, 188

• *F* •

face on Mars, 103
festoon, defined, 126
fireball meteor, 59–60, 302
flashlight, using red-light, 47, 63
following spot (sunspots), 151
Ford, Kent (dark matter studies), 246–247
fossil evidence, of life on Mars, 103
Foucault pendulum, 42–43
full Moon, 88, 93
full-aperture solar filters, 159

• *G* •

galactic bulge, 200, 230
galactic center, of Milky Way, 200
galactic disk, of Milky Way, 199
galactic equator, of Milky Way, 200
Galactic Latitude, 200
Galactic Longitude, 200
galactic plane, of Milky Way, 200
galactic rim, of Milky Way, 201
galactic year, 27, 200
galaxies. *See also specific galaxies*
   clusters of, 217–218
   dark matter effect on, 246–247
   defined, 302
   Local Group of, 216–217
   low surface brightness, 214
   versus nebulae, 208
   radio, 228
   Seyfert, 229
   types of, 211–214
   viewing, 214–216
Galilean moons, 127–130
Galilei, Galileo (17th-century Italian
   astronomer)
   projection technique of, 156
   stopping down of telescopes, 159
   viewing sun, 155
gamma ray bursts, 191–192, 302
Gamow, George (physicist), 254
Ganymede (Jupiter's moon), 128, 129
gas-giant planets, 124. *See also* Jupiter;
   Neptune; Saturn; Uranus
Geminids (meteor showers), 62

General Theory of Relativity (Albert Einstein), 27
geography
of Mars, 101–102
of Mercury, 98–99
of Venus, 99–100
geology, of moon, 90–91
geomagnetic storms, solar wind and, 153
Georgia State University Observatory, 35
Giant Impact theory, moon's origin, 94–95
giant molecular clouds, 206
globular star clusters, 189, 203–205
glow aurora, 79
gravitational lensing, 193, 250–251
gravity
about, 26–27
of black holes, 176–177
effect on sun, 147
shape of Milky Way and, 199
Great Walls, 197, 218
greatest eastern elongation, 106, 108
greatest western elongation, 106, 108
Greek alphabet
letters and corresponding symbols, 14
naming stars and, 13
Greenwich Mean Time, defined, 83
Griffith Observatory and Planetarium, 35
Guth, Allen (inflation theory), 255

• H •

H II regions, 171, 205
Hale-Bopp, comet, 69
half Moon, 88
Halley's Comet, 65, 69
halo event (sun), 150
H-alpha solar filters, 158
Hayden Planetarium, 36
helium, Big Bang theory and, 255
Herbig-Haro objects, 171
Hertzsprung, Elnar (H-R diagram), 178
Hertzsprung-Russell diagram. See H-R diagram
Hipparchos (developing magnitude classes), 20–21
horizon, naked-eye observation and, 47
hot Jupiters (extrasolar planets), 242–243
H-R diagram
brown dwarf stars on, 183
classifying luminosity on, 180
interpreting, 182
plotting spectral type, 178–179
spectral classes for, 179–180
star mass and, 181
Hubble constant, 261
Hubble, Edwin P. (astronomer)
expanding universe discoveries, 254
The Realm of the Nebulae, 208
Hubble Flow, 27
Hubble Space Telescope, 73, 74, 271
Hubble's Law, 261
Hyades star cluster, 202
hydrogen-burning shell, of red giants, 172
hydrosphere, 80
Hygiea (asteroid), 116
hypernovas, 191–192

• I •

ice, on Mars, 101
icy planets, 136
impact basins, on Mercury, 98
impact craters, 90, 98
inferior conjunction, 107, 108
inferior planet, 107
inflation
Big Bang theory and, 255–256
energy powering, 256–257
shape of universe and, 257
interarm regions, of Milky Way, 199
intermediate mass black holes, 178
International Astronomical Union
naming sky objects and, 71
purchased star names and, 15
reporting comets to, 72
standards by, 13
International MarsWatch 2003 Web site, monitoring Mars and, 112
International Space Station, seeing, 73, 74
Internet
astronomy resources on, 31–32
finding astronomy clubs on, 30–31
observatories on, 35–36
satellite viewing predictions on, 74
viewing sun images on, 167

intrinsic variable stars, 188
Io (Jupiter's moon), 127, 128
Iridium satellites, viewing, 74
irregular galaxies, 213

• *J* •

jets, of quasars, 226
Jewel Box star cluster, 203
Jewitt, David (finding Kuiper Belt
    Objects), 139
Jupiter
    about, 123–125
    comets hitting, 129
    Great Red Spot on, 126
    moons of, 127–130
    rings of, 127

• *K* •

Kirshner, Robert (Bootes Void), 218
Kitt Peak National Observatory, 35
Kuiper Belt Objects (KBOs), 139–140

• *L* •

Lagoon Nebula, 210
lame duck stars, 22
Large Magellanic Cloud Galaxy, 201,
    215–216
leading spot (sunspots), 151
Leavitt, Henrietta (astronomer), 188
lenticular galaxies, 212
Leonids (meteor showers), about, 62
Levy, David (amateur astronomer), 70, 129
life
    on Earth, 78, 79
    on Mars, 102–103
light, distinguishing objects with, 11–12
light pollution
    interfering with stargazing, 46–47
    viewing in city versus remote areas, 9
    viewing Milky Way and, 198
light-years
    defined, 304
    measuring distance with, 22–23, 269
limiting magnitude, 47
Lincoln Near Earth Asteroid Research
    (LINEAR) project, 120

lithosphere, 80
Little Dipper, finding North Star in, 44
Local Apparent Sidereal Time, 84
Local Group of Galaxies
    about, 216–217
    Earth's movement around, 27
Loch Ness Productions planetariums, 36
long period variable stars, 189
low surface brightness galaxies, 214
Lowell Observatory, 34–35
Lowell, Percival (astronomer)
    Mars canals theory, 102
    theory of Pluto's existence, 267
lucida, identifying stars as, 15
luminosity
    classifying stars, 180
    of sun, 146
lunar charts, sources of, 92
lunar eclipse, 89–90
lunar highlands, 91
lunar limb, 163
lunar mountains, 92
Lunar Prospector
    magnetic fields on moon and, 95
    water on moon and, 91
lunation, moon phases and, 88
Luu, Jane (finding KBOs), 139
Lyra constellation, stars in, 195
Lyrids (meteor showers), 62

• *M* •

MACHOs (massive compact halo objects),
    searching for, 250
magazines, astronomy resources, 32–33
magnetic fields
    on Earth, 81
    on Jupiter and Saturn, 124
    on Mars, 101
    on Mercury, 98
    on moon, 95
    on sun, 149–150
magnetographs, 149–150
magnetosphere
    defined, 81
    solar plasma and, 153
magnitude
    of brightest stars, 46
    defined, 304

definitions for differing purposes of, 47
measuring star, 16, 20–22
systematic progression of, 21
main sequence stars
about, 172
on H-R diagram, 182
star lifecycle and, 170
Maksutov-Cassegrain telescopes
about, 51
advantage of, 54
sun viewing with, 156
Marcy, Geoff (extrasolar planets), 243
Maria Mitchell Observatory, 35
maria, on moon, 91
Mariner 10 spacecraft, Mercury and, 97–98
Mars
about, 101–102
backtracking of, 111
comparing to Earth, 104
investigations of, 97
rocks from, on Earth, 256–267
theories of life on, 102–103
viewing, 110–113
Mars Global Surveyor (MGS), 100
Mars Odyssey spacecraft, 100
massive compact halo objects (MACHOs),
searching for, 250
Mauna Kea Observatories, 35
maxima, 193
Mayor, Michel (extrasolar planets), 242
Meade ETX-90PE telescope
about, 54–55
sun viewing with, 156
Merak (star), finding North Star with, 44
Mercury
about, 98–99
comparing to Earth, 103–104
investigations of, 97–98
transit of, 113
viewing, 105, 107, 113–114
MESSENGER (MErcury Surface, Space
ENvironment, GEochemistry, and
Ranging), probing Mercury, 98
Messier 32 Galaxy, viewing, 215
Messier Catalog, 20
Messier Catalog Web site of Students for
the Exploration and Development of
Space, 20

Messier, Charles (18th-century French
astronomer), 20
The Messier Objects (Stephen J.
O'Meara), 20
meteor crater, 118
meteor showers
about, 59, 61–63
defined, 61
photographing, 64
top annual, 62
viewing, 63–64
meteor train, 60
meteorites
about, 58
defined, 57, 302
naming, 71
radioactive dating, 86
temperature of, 269
meteoroids
defined, 57, 302
types of, 58
meteors
about, 59–61
versus comets, 65
defined, 57, 302
naming, 71
photographing, 64
MGS (Mars Global Surveyor), 100
microlensing, defined, 193
microlensing events, viewing, 193
micrometeorites
about, 58
on Earth, 265
Milky Way
about, 197–198
history of, 198–199
locating, 200–201
naked-eye observation and, 46
shape of, 199–200
viewing objects beyond, 201
minima, 193
Minor Planet Center (MPC), 118
Mira stars, 189
moon (Earth's)
about, 87
asteroids hitting, 117
dark side of, 93–94
geology of, 90–91
lunar eclipse, 89–90

moon (Earth's) *(continued)*
  phases of, 87–89
  theory of origin of, 94–95
  viewing near side of, 91–93
moon maps, sources of, 92
moon shadow, on Jupiter, 128–129
moons
  Jupiter's, 127–130, 133
  Neptune's, 137
  Pluto's, 138–139
  Saturn's, 132–133
  types of, 133
  Uranus's, 136–137
Morning Star
  changing, 105
  planets as, 270
motion, and space, 27–28
Mount Wilson Observatory, 35
mounts, for telescopes, 52–53
MPC (Minor Planet Center), 118
multiple stars, 187

• *N* •

naked-eye observation
  about, 45–47
  brightest stars for, 46
  purpose of, 41
NASA, Web site as scientific resource, 32.
    See also specific NASA undertakings
National Optical Astronomy
    Observatory, 35
National Radio Astronomy Observatory, 35
National Solar Observatory, 35
navigation, star, 11
NCP (North Celestial Pole)
  reading star maps and, 25
  stargazing from Northern Hemisphere, 42
Near Earth Objects (NEOs)
  about, 118–119
  defined, 302
  protecting Earth from, 119–120
Near-Earth Asteroid Tracking (NEAT)
    project, 120
nebulae
  defined, 205, 302
  familiar, 205–206
  versus galaxies, 208
  planetary, 206–207

star lifecycle and, 170
  sun becoming, 155
  supernova remnants as, 190–191
  viewing, 208–210
Neptune
  about, 135–136, 137
  viewing, 141–142
neutrinos
  defined, 302
  sun releasing, 153–154
neutron stars, 175–176, 302
new crescent Moon, 88
new Moon, 88
Newton, Isaac (concept of gravity), 26
Newtonian reflector telescopes
  about, 51
  example of, 52
  projection technique with, 156–158
NGC 205 Galaxy, viewing, 215
NGC 6231 star cluster, 203
Night Sky magazine, 32
North American Meteor Network, 59
North American Nebula, 209
North Celestial Pole (NCP)
  reading star maps and, 25
  stargazing from Northern Hemisphere, 42
North Star (Polaris)
  finding, 43–44
  as a reference point, 42
  tilt of Earth's axis and, 84
Northern Coal Sack nebula, 209
Northern Hemisphere
  change of seasons in, 84–86
  star maps for, 293, 295, 297, 299
Northern Lights (aurora borealis), 79
*Norton's Star Atlas and Reference
    Handbook*, viewing Mars with, 112
Nova Search program, 195–196
novas, 190
nuclear fusion, sun's energy and, 146
nucleus, of comets, 65

• *O* •

OB associations, 205
observatories, using, 34–36
Observer's Handbook of the Royal
    Astronomical Society of Canada, 33

occultation
  defined, 121, 302
  timing asteroidal, 121–122
  tracking asteroidal, 122
  viewing Jupiter's moons and, 128
occultation ground track, 122
oddball dark matter, 248–249
off-axis solar filters, 159
Olympus Mons (Mars volcano), 101–102
O'Meara, Stephen J. (*The Messier Objects*), 20
open star clusters, 202–203
opposition, 106, 107
optical double stars, 187
optically violently variable quasars (OVVs), 228
orbit
  of binary stars, 183–184
  defined, 302
  Earth, seasons and, 84–86
  Earth, time and, 82–83
  Mars, seasons and, 101
  moon phases and, 88–89
  of multiple stars, 187
  Newton's law of gravity and, 26
  prograde and retrograde, 133
  stars, dark matter and, 246–247
  superior versus inferior planets, 107
Orion constellation
  about, 44–45
  stars in, 195
Orion Nebula
  about, 209
  image of, 171
Orion ShortTube 80 mm refractor telescope, 70
Orion Telescopes & Binoculars, source of moon maps/lunar charts, 92
Orionids (meteor showers), 62
OVVs (optically violently variable quasars), 228
oxygen, on Earth, 78

● *P* ●

Pallas (asteroid), 116
Palomar Observatory, 35
parent atoms, radioactive dating and, 86
parsec, 217

path of totality
  defined, 38
  of solar eclipse, 165
  viewing lunar eclipses and, 90
Penzias, Arno (cosmic microwave background), 254–255
period-luminosity relation, of Cepheid variable stars, 188–189
Perseids (meteor showers), 61–62
Personal Solar Telescope, 158
PHAs (Potentially Hazardous Asteroids), 118–119
photography
  color enhanced, 53
  meteors/meteor showers and, 64
  solar, 167
photosphere, 147, 148–149
planet position, understanding, 106–108
planetarium programs
  about, 33–34
  acquiring basic, 56
  searching for comets and, 71
planetariums, 34, 36
planetary nebulae, 206–207, 303
planets. *See also specific planets*
  about, 12
  defined, 302
  finding extrasolar, 240–244
  as Morning Star, 270
  movement across sky, 43
  visible at dawn 2006–2010, 277–278, 280–282, 284–285, 286–288, 290–291
  visible at dusk 2006–2010, 276–277, 279–280, 282–283, 285–286, 288–290
planisphere, tracking constellations with, 43
plasma tail, of comets, 67–68
plate tectonics
  on Earth, 78
  on Venus, 100
Pleiades (Seven Sisters) star cluster, 202
Plutinos, 140
Pluto
  about, 137–138
  asteroid theory of, 139
  KBOs and, 139–140
  moon of, 138–139
  predictions of existence of, 267
  viewing, 142

polar ice, on Mars, 101
Polaris. *See* North Star (Polaris)
polarities, sunspot cycle and, 152
Ponticus, Heraclides (Greek
        philosopher), 42
position, measuring star, 24–26
Potentially Hazardous Asteroids (PHAs),
        118–119
prograde orbits, 133
Project Phoenix, 237–238
projection technique, 156–158
prominences (sun), 149
protoplanetary nebulae, 207
Proxima Centauri
    Alpha Centauri system and, 193–194
    distance of, 22
publications, astronomy resources, 32–33.
        See also specific publications
pulsars, 175–176, 303

● *Q* ●

Quadrantids (meteor showers), 62
quadruple stars, 187
quantum fluctuations, universe and, 256
quarter Moon, 88
quasars. *See also* active galactic nuclei
        (AGN)
    defined, 225–226, 303
    measuring size of, 226
    spectra of, 226–227
    types of, 227–228
quasi-stellar objects (QSOs), 227
Queloz, Didier (extrasolar planets), 242

● *R* ●

RA (right ascension)
    defined, 106, 304
    measuring star positions, 23–26
radiant, of meteor showers, 61
radiation. *See* cosmic microwave
        background
radio astronomy observatories, 35
radio galaxies, 228

radio telescope
    Allen Telescope Array, 239
    image of, 235
    search for aliens and, 234
radioactive dating, measuring Earth's age
        with, 86
radio-loud quasars, 227
radio-quiet quasars, 227
rays aurora, 79
rays, viewing lunar, 92
The Realm of the Nebulae (Edwin P.
        Hubble), 208
red dwarfs, 172, 190
red giants, 170, 172, 303
red supergiants, 172, 182
redshift, 224, 303
reflection nebulae, 206
reflector telescopes. *See also* telescopes
    about, 51
    example of, 52
    projection technique with, 156–158
    versus refractor, 54
refractor telescopes. *See also* telescopes
    about, 51
    projection technique with, 156–158
    versus reflector, 54
retrograde orbits, 133
revolving, defined, 27
Rhea (Saturn's moon), 133
ridges, on Mercury, 98
Rigel (star), in Orion, 44–45
right ascension (RA)
    defined, 106, 304
    measuring star positions, 24–26
rilles
    defined, 91
    on moon, 92
    on Venus, 99–100
Ring Nebula, 209
rings
    of Jupiter, 127
    of Saturn, 130–132
    of Uranus, 136–137
rotation
    defined, 27, 303
    Earth's changing sky and, 43

Earth's, measuring time and, 82–84
Earth's, proof of, 42–43
Jupiter's, 124
Saturn's, 131
synchronous, 93
of Uranus and Neptune, 136
Royal Astronomical Society of Canada, 31
Royal Astronomical Society of Canada
Observer's Handbook, 33
RR Lyrae stars, 189
Rubin, Vera (dark matter studies), 246–247
Russell, Henry Norris (H-R diagram), 178

• *S* •

Sagittarius A*
about, 177
in galactic bulge, 200
Sagittarius Dwarf Galaxy, 215
satellites. *See* artificial satellites; moons
Saturn
about, 123–124, 130
rings of, 130–132
storms on, 132
Schmidt-Cassegrain telescopes
about, 51
advantage of, 54
sun viewing with, 156
scientific data, amateur astronomers
gathering, 39
SCP (South Celestial Pole), reading star
maps and, 25
Sculptor Galaxy, viewing, 216
seafloor, magnetic properties of Earth's, 82
Search for Extraterrestrial Intelligence
(SETI)
defined, 303
joining projects of, 240
listening criteria of, 235–237
programs, 238–240
Project Phoenix, 237–238
using Drake Equation, 234
Search for Extraterrestrial Radio Emissions
from Nearby Developed Intelligent
Populations (SERENDIP), 238
seasons
cause of Earth's, 84–86
Earth's distance from sun and, 270
on Mars, 101

seeing
defined, 303
good versus bad, 55
sky conditions and, 112
SETI (Search for Extraterrestrial
Intelligence)
defined, 303
joining projects of, 240
listening criteria of, 235–237
programs, 238–240
Project Phoenix, 237–238
using Drake Equation, 234
Seyfert galaxies, 229
shadow bands, during eclipse, 164–165
sidereal clocks, use of, 83
sidereal day, 83
singularity, of black holes, 222–223
Sirius (dog star), 194–195
site, ideal stargazing, 47
sky conditions, rating, 112
sky darkness, 112
sky maps. *See* star maps
sky objects
artificially coloring, 53
distinguishing, light and, 11–12
naming, 71
Sky & Space magazine, 33
Sky & Telescope Australia magazine, 33
Sky & Telescope magazine Web site, 32
*SkyNews: The Canadian Magazine of
Astronomy & Stargazing*, 32
Small Magellanic Cloud Galaxy
near Milky Way, 201
viewing, 215–216
solar activity
about, 149–152
defined, 303
Solar and Heliospheric Observatory
(SOHO) satellite, 167
solar constant, 152
solar eclipses
about, 162
safely viewing, 163–164
table of future, 166
solar filters
front-end, 159–160
H-alpha, 158
sun safety and, 54
viewing Eclipse without, 163–164

solar flares
  about, 150–151
  creating solar wind, 153
solar luminosity, 152
solar photography, 167
solar plasma, 153
solar system, Earth and, 77
solar wind
  about, 153
  plasma tails and, 67
Sombrero Galaxy, viewing, 215
South Celestial Pole (SCP), reading star
      maps and, 25
Southern Hemisphere
  change of seasons in, 84–86
  star maps for, 294, 296, 298, 300
Southern Lights (aurora australis), 79
Southern SERENDIP, 238
space
  motion and, 27–28
  warped by black holes, 224–225
space objects. *See* sky objects
Spaceguard Foundation, 120
spectral classes, of stars, 179
spectral type, 178–179, 303
spectrum
  defined, 178
  quasar, 226–227
spiral galaxies
  about, 211–212
  dark matter effect on, 246–247
sporadic meteor, 59
standard candle, measuring with, 262
standard time zones, converting to
      Universal Time, 84
star clusters
  about, 201
  defined, 303
  globular, 203–205
  open, 202–203
star maps
  acquiring basic, 56
  comparison charts, 195
  International Astronomical Union and, 13
  labeling stars on, 15
  Northern Hemisphere, 293, 295, 297, 299
  Southern Hemisphere, 294, 296, 298, 300
  tracking constellations with, 43
  using RA and Dec and, 25

star parties, 37
Stark Effect, understanding, 186
stars
  Alpha Centauri system, 193–194
  analyzing lines in spectra of, 186
  ancient understanding of, 11
  artificially coloring, 53
  binary, 183–185
  brightest, 45–46
  brightest, in constellations, 16–19
  classifying luminosity of, 180
  dark matter effect on, 246–247
  defined, 303
  distinguishing, 11–12
  end states of stellar evolution, 170,
      173–178
  exploding, 190–192
  exploding, invisibility of, 267
  flare, 190
  lifecycle of, 169–170
  main sequence, 170, 172, 182
  mass determining class of, 181
  massive, 173
  measuring distance of, 22–23
  multiple, 187
  naming, 12–16
  neighboring Earth, 193–195
  pulsating, 188–189
  spectral classes for, 179–180
  spectral type of, 178–179
  sun as, 146
  variable, 187–188
  viewing with comparison charts, 195
  young stellar objects, 170, 171, 181
stellar interior, 147
stellar mass black holes, 177, 220
stellar spectroscopy, 186
Sterne und Weltraum magazine, 33
sun
  about, 145–146
  gravity and, 147
  life expectancy of, 154–155
  movement across sky, 43
  regions of, 147–149
  releasing neutrinos, 153–154
  safely viewing, 54, 145, 155–160
  versus other stars, 271
  viewing Internet images, 167
  viewing with H-alpha solar filters, 158

sun's magnetic cycle, 152
sunspot cycle
  defined, 151
  polarities and, 152
sunspot numbers, 161–162
sunspots
  about, 151–152
  darkness of, 267
  tracking, 160–162
superclusters, 197, 218
superior conjunction, 107, 108
superior planet, 107
superluminal motion, of quasar jets, 226
supermassive black holes, 177, 220
supernova remnants, 190–191, 208
Supernova Search program, 196
supernovas
  about, 174–175, 190–191
  defined, 303
  in Large Magellanic Cloud galaxy, 201
Swift satellite, hypernovas and, 191–192
symbols, Greek alphabet, 14
synchronous rotation, 93
Syrtis Major (Mars), 111
systematic progression, star magnitude, 21

• T •

T Tauri stars, 171
tails, comet, 67–68, 265
Tarantula Nebula, 201, 210
telescope motels, 40
telescopes
  about, 50
  acquiring basic, 56
  for comet searching, 70
  distinguishing lunar elements with, 92
  front-end solar filters for, 159–160
  H-alpha, 158
  mounts for, 52–53
  projection technique with, 156–158
  purpose of, 42
  radio, 234, 235, 239
  selecting, 53–55
  sun safety and, 54, 156–160
  understanding classifications, 51–52
  warm versus cool, 55

temperatures
  on Jupiter and Saturn, 124
  on Mars, 101
  on Mercury, 99
  on moon, 93
  on Venus, 99
terminator
  defined, 92, 304
  on Venus, 108
  viewing moon and, 92–93
terrestrial planet, 77
time
  measuring, Earth's rotation and, 82–83
  understanding standard systems of,
    83–84
  warped by black holes, 224–225
Titan (Saturn's moon), 132
Tombaugh, Clyde (finding Pluto), 139, 267
topographical mapping, of Mars, 100
tours, eclipse, 38, 39–40
transit
  defined, 304
  of Mercury, 113
  of Venus, 110
  viewing Jupiter's moons and, 128
transit technique, 241
transition region (sun), 149
transparency (sky conditions), 112
Triangulum Galaxy
  near Milky Way, 201
  viewing, 215
Trifid Nebula, 210
triple stars, 187
Triton (Neptune's moon), 137
turbulence, good versus bad seeing and, 55

• U •

Unified Model of Active Galactic
    Nuclei, 230
Universal Time (UT), 83, 84
universe
  age of, 260–261
  effect of quantum fluctuations on, 256
  expanding, Big Bang theory and, 254
  mapping, 260
  shape of, critical density and, 257
  understanding, 258–259

Uranus
  about, 135–137
  viewing, 140–141
Ursa Major, using Big Dipper from, 44
U.S. Naval Observatory (USNO), 84
UT (Universal Time), 83, 84
UTC (Coordinated Universal Time), 83, 84

• *V* •

vacations, astronomy, 36–40
vacuum, powering inflation, 256–257
Valles Marineris (Mariner Valley), 102
variable stars
  about, 187–188
  defined, 304
  pulsating, 188–189
Vega, in Lyra constellation, 195
Venus
  about, 99–100
  comparing to Earth, 104
  as Evening and Morning Star, 105
  fireball meteors and, 59
  investigations of, 97
  rain on, 99, 266
  transit of, 110
  viewing, 104–105, 107, 108–110
Venus lander spacecraft, 99
vernal equinox, 85
Vesta (asteroid), 116, 117
virga rain, on Venus, 99, 266
volcanism
  defined, 86
  on Earth, 78
  on Mars, 101–102
  on moon, 91
  on Venus, 100

• *W* •

waning crescent Moon, 88–89
waning gibbous Moon, 88

water
  on Earth, 78
  on Mars, 101
  on Uranus and Neptune, 136
waxing crescent Moon, 88
waxing gibbous Moon, 88
weakly interacting massive particles
    (WIMPs), searching for, 249–250
weather. *See* temperatures
Web sites. *See* Internet
Whirlpool Galaxy, viewing, 215
white dwarfs
  about, 174
  defined, 304
  on H-R diagram, 182
  star lifecycle and, 170
  sun becoming, 155
white hole, in black holes, 222–223
Wilkinson Microwave Anisotropy Probe
    (WMAP), 260
Wilson, Robert (cosmic microwave
    background), 254–255
WIMPs (weakly interacting massive
    particles), searching for, 249–250
wormhole, in black holes, 222

• *Y* •

Young Stellar Objects (YSOs)
  about, 171
  plotted on H-R diagrams, 181
  star lifecycle and, 170

• *Z* •

Zeeman Effect, understanding, 186
Zenith, 64, 304

BUSINESS, CAREERS & PERSONAL FINANCE

0-7645-5307-0

0-7645-5331-3 *†

Also available:

- Accounting For Dummies †
  0-7645-5314-3
- Business Plans Kit For Dummies †
  0-7645-5365-8
- Cover Letters For Dummies
  0-7645-5224-4
- Frugal Living For Dummies
  0-7645-5403-4
- Leadership For Dummies
  0-7645-5176-0
- Managing For Dummies
  0-7645-1771-6

- Marketing For Dummies
  0-7645-5600-2
- Personal Finance For Dummies *
  0-7645-2590-5
- Project Management For Dummies
  0-7645-5283-X
- Resumes For Dummies †
  0-7645-5471-9
- Selling For Dummies
  0-7645-5363-1
- Small Business Kit For Dummies *†
  0-7645-5093-4

HOME & BUSINESS COMPUTER BASICS

0-7645-4074-2

0-7645-3758-X

Also available:

- ACT! 6 For Dummies
  0-7645-2645-6
- iLife '04 All-in-One Desk Reference
  For Dummies
  0-7645-7347-0
- iPAQ For Dummies
  0-7645-6769-1
- Mac OS X Panther Timesaving
  Techniques For Dummies
  0-7645-5812-9
- Macs For Dummies
  0-7645-5656-8

- Microsoft Money 2004 For Dummies
  0-7645-4195-1
- Office 2003 All-in-One Desk Reference
  For Dummies
  0-7645-3883-7
- Outlook 2003 For Dummies
  0-7645-3759-8
- PCs For Dummies
  0-7645-4074-2
- TiVo For Dummies
  0-7645-6923-6
- Upgrading and Fixing PCs For Dummies
  0-7645-1665-5
- Windows XP Timesaving Techniques
  For Dummies
  0-7645-3748-2

FOOD, HOME, GARDEN, HOBBIES, MUSIC & PETS

0-7645-5295-3

0-7645-5232-5

Also available:

- Bass Guitar For Dummies
  0-7645-2487-9
- Diabetes Cookbook For Dummies
  0-7645-5230-9
- Gardening For Dummies *
  0-7645-5130-2
- Guitar For Dummies
  0-7645-5106-X
- Holiday Decorating For Dummies
  0-7645-2570-0
- Home Improvement All-in-One
  For Dummies
  0-7645-5680-0

- Knitting For Dummies
  0-7645-5395-X
- Piano For Dummies
  0-7645-5105-1
- Puppies For Dummies
  0-7645-5255-4
- Scrapbooking For Dummies
  0-7645-7208-3
- Senior Dogs For Dummies
  0-7645-5818-8
- Singing For Dummies
  0-7645-2475-5
- 30-Minute Meals For Dummies
  0-7645-2589-1

INTERNET & DIGITAL MEDIA

0-7645-1664-7

0-7645-6924-4

Also available:

- 2005 Online Shopping Directory
  For Dummies
  0-7645-7495-7
- CD & DVD Recording For Dummies
  0-7645-5956-7
- eBay For Dummies
  0-7645-5654-1
- Fighting Spam For Dummies
  0-7645-5965-6
- Genealogy Online For Dummies
  0-7645-5964-8
- Google For Dummies
  0-7645-4420-9

- Home Recording For Musicians
  For Dummies
  0-7645-1634-5
- The Internet For Dummies
  0-7645-4173-0
- iPod & iTunes For Dummies
  0-7645-7772-7
- Preventing Identity Theft For Dummies
  0-7645-7336-5
- Pro Tools All-in-One Desk Reference
  For Dummies
  0-7645-5714-9
- Roxio Easy Media Creator For Dummies
  0-7645-7131-1

* Separate Canadian edition also available
† Separate U.K. edition also available

Available wherever books are sold. For more information or to order direct: U.S. customers visit www.dummies.com or call 1-877-762-2974. U.K. customers visit www.wileyeurope.com or call 0800 243407. Canadian customers visit www.wiley.ca or call 1-800-567-4797.

 **WILEY**

SPORTS, FITNESS, PARENTING, RELIGION & SPIRITUALITY

0-7645-5146-9

0-7645-5418-2

Also available:

- Adoption For Dummies
  0-7645-5488-3
- Basketball For Dummies
  0-7645-5248-1
- The Bible For Dummies
  0-7645-5296-1
- Buddhism For Dummies
  0-7645-5359-3
- Catholicism For Dummies
  0-7645-5391-7
- Hockey For Dummies
  0-7645-5228-7

- Judaism For Dummies
  0-7645-5299-6
- Martial Arts For Dummies
  0-7645-5358-5
- Pilates For Dummies
  0-7645-5397-6
- Religion For Dummies
  0-7645-5264-3
- Teaching Kids to Read For Dummies
  0-7645-4043-2
- Weight Training For Dummies
  0-7645-5168-X
- Yoga For Dummies
  0-7645-5117-5

TRAVEL

0-7645-5438-7

0-7645-5453-0

Also available:

- Alaska For Dummies
  0-7645-1761-9
- Arizona For Dummies
  0-7645-6938-4
- Cancún and the Yucatán For Dummies
  0-7645-2437-2
- Cruise Vacations For Dummies
  0-7645-6941-4
- Europe For Dummies
  0-7645-5456-5
- Ireland For Dummies
  0-7645-5455-7

- Las Vegas For Dummies
  0-7645-5448-4
- London For Dummies
  0-7645-4277-X
- New York City For Dummies
  0-7645-6945-7
- Paris For Dummies
  0-7645-5494-8
- RV Vacations For Dummies
  0-7645-5443-3
- Walt Disney World & Orlando For Dummies
  0-7645-6943-0

GRAPHICS, DESIGN & WEB DEVELOPMENT

0-7645-4345-8

0-7645-5589-8

Also available:

- Adobe Acrobat 6 PDF For Dummies
  0-7645-3760-1
- Building a Web Site For Dummies
  0-7645-7144-3
- Dreamweaver MX 2004 For Dummies
  0-7645-4342-3
- FrontPage 2003 For Dummies
  0-7645-3882-9
- HTML 4 For Dummies
  0-7645-1995-6
- Illustrator CS For Dummies
  0-7645-4084-X

- Macromedia Flash MX 2004 For Dummies
  0-7645-4358-X
- Photoshop 7 All-in-One Desk Reference For Dummies
  0-7645-1667-1
- Photoshop CS Timesaving Techniques For Dummies
  0-7645-6782-9
- PHP 5 For Dummies
  0-7645-4166-8
- PowerPoint 2003 For Dummies
  0-7645-3908-6
- QuarkXPress 6 For Dummies
  0-7645-2593-X

NETWORKING, SECURITY, PROGRAMMING & DATABASES

0-7645-6852-3

0-7645-5784-X

Also available:

- A+ Certification For Dummies
  0-7645-4187-0
- Access 2003 All-in-One Desk Reference For Dummies
  0-7645-3988-4
- Beginning Programming For Dummies
  0-7645-4997-9
- C For Dummies
  0-7645-7068-4
- Firewalls For Dummies
  0-7645-4048-3
- Home Networking For Dummies
  0-7645-42796

- Network Security For Dummies
  0-7645-1679-5
- Networking For Dummies
  0-7645-1677-9
- TCP/IP For Dummies
  0-7645-1760-0
- VBA For Dummies
  0-7645-3989-2
- Wireless All In-One Desk Reference For Dummies
  0-7645-7496-5
- Wireless Home Networking For Dummies
  0-7645-3910-8

Astronomy For Dummies, 2nd Edition

The Space Age

1957 The Soviet Union launches Sputnik 1, the first artificial satellite to orbit Earth; Geoffrey Burbidge, E. Margaret Burbidge, William Fowler, and Fred Hoyle explain how elements form in stars.

1958 Using the satellite Explorer 1, James Van Allen discovers Earth's radiation belts (*magnetosphere*).

1960 Frank Drake begins the Search for Extra-terrestrial Intelligence (SETI) at the National Radio Astronomy Observatory in Green Bank, West Virginia.

1961 Yuri Gagarin makes the first manned space flight.

1963 Valentina Tereshkova is the first woman in space.

1967 Jocelyn Bell Burnell and Anthony Hewish discover pulsars.

1969 Neil Armstrong and Buzz Aldrin walk on the moon.

1979 Using pictures from Voyager 1, Linda Morabito discovers erupting volcanoes on Jupiter's moon Io.

1987 Ian Shelton discovers the first supernova, since 1604, plainly visible to the naked eye.

1990 The Hubble Space Telescope launches.

1991 Alexander Wolszczan discovers planets orbiting a pulsar — the first known planets outside the solar system.

1995 Michel Mayor and Didier Queloz discover 51 Pegasi B, the first planet of a normal star beyond the sun.

1998 Two astronomer teams discover that the expansion of the universe is getting faster, perhaps due to a mysterious force associated with the vacuum of space.

1999 Mars Global Surveyor finds that Mars may have had an ocean at one time.

2003 The Wilkinson Microwave Anisotropy Probe satellite finds that the universe is 13.7 billion years old.

2005 The Huygens space probe lands on Titan, Saturn's largest moon.

Famous Women in Astronomy

Historical:

Caroline Herschel (1750–1848) Discovered eight comets.

Annie Jump Cannon (1863–1941) Devised the basic method for classifying the stars.

Henrietta Swan Leavitt (1868–1921) Discovered the first accurate method for measuring great distances in space.

Contemporary:

E. Margaret Burbidge Pioneered modern studies of galaxies and quasars.

Jocelyn Bell Burnell Discovered pulsars in her work as a graduate student.

Wendy Freedman Leader in measuring the expansion rate of the universe.

Carolyn C. Porco Leads the Cassini imaging science team in the study of Saturn and its moons and rings.

Sally Ride A trained astrophysicist and the first American woman in space.

Nancy G. Roman As NASA's first chief astronomer, she led the development of telescopes in space.

Vera C. Rubin Investigated the rotation of galaxies and detected the existence of dark matter.

Carolyn Shoemaker Discovered many comets, including one that smashed into Jupiter.

Jill Tarter Leader of the largest search for extraterrestrial intelligence, Project Phoenix.

Astronomy For Dummies, 2nd Edition

An Astronomical Timeline

2000 B.C. According to legend, two Chinese astronomers are executed for not predicting an eclipse and for being drunk as it happened.

129 B.C. Hipparchos completes the first catalog of the stars.

A.D. 150 Ptolemy publishes his theory of the Earth-centered universe.

970 al-Sufi prepares catalog of over 1,000 stars.

1420 Ulugh-Beg, prince of Turkestan, builds a great observatory and prepares tables of planet and star data.

1543 While on his deathbed, Copernicus publishes his theory that planets orbit around the sun.

1609 Galileo discovers craters on Earth's moon, the moons of Jupiter, the turning of the sun, and the presence of innumerable stars in the Milky Way with a telescope that he built.

1666 Isaac Newton begins his work on the theory of universal gravitation.

1671 Newton demonstrates his invention, the reflecting telescope.

1705 Edmond Halley predicts that a great comet will return in 1758.

1758 On Christmas, farmer/amateur astronomer Johann Palitzch discovers the return of Halley's Comet.

1781 William Herschel discovers Uranus.

1791 Benjamin Banneker, the first African American scientist, begins star observations needed for the geographical survey to establish the future capital city of the United States, Washington, D.C.

1833 Abraham Lincoln and thousands of others see an enormous meteor shower over North America on November 12 and 13.

1842 Christian Doppler discovers the principle by which sound or light shifts in frequency and wavelength due to the motion of its source with respect to the observer.

1846 Johann Galle is the first person to spot Neptune.

1910 Earth passes through the tail of Halley's Comet.

1916 Albert Einstein proposes the General Theory of Relativity, which explains the nature of gravity and the bending of light as it passes the sun, predicts the existence of black holes, and details the twisting of time and space in the vicinity of a massive, spinning object.

1923 Edwin Hubble proves that other galaxies lie beyond the Milky Way.

1926 The first launch of a liquid-fuel rocket, developed by Robert Goddard.

1930 Clyde Tombaugh discovers Pluto.

1931 Karl Jansky discovers radio waves from space.

1939 Hans Bethe explains the energy source of the sun and other stars.

1940 Grote Reber reports the first radio telescope survey of the sky.

For Dummies: Bestselling Book Series for Beginners